The Marathon of the Messenger

WE HOPE THIS BOOK WILL INOCULATE YOU AGAINST MISCONCEPTIONS ABOUT MRNA VACCINE...
AND WILL ACT AS A SANE ALTERNATIVE!
JÉRÔME LEMONNIER
NICOLAS LEMONNIER
CHARROT.

Jérôme Lemonnier · Nicolas Lemonnier

The Marathon of the Messenger

A History of Messenger RNA Vaccines

Drawings by Gilles Charrot
Foreword by Pierre Meulien

 Springer

Jérôme Lemonnier
Wasquehal, France

Nicolas Lemonnier
Orsay, France

Drawings by
Gilles Charrot
Montrouge, France

With Contributed by
Chantal Pichon
UPR4301 CNRS
Centre de biophysique moléculaire
Orléans, France

Steve Pascolo
Department of Dermatology
University Hospital Zürich
Zürich, Switzerland

ISBN 978-3-031-39299-3   ISBN 978-3-031-39300-6   (eBook)
https://doi.org/10.1007/978-3-031-39300-6

The translation was done with the help of artificial intelligence (machine translation by the service DeepL. com). A subsequent human revision was done primarily in terms of content.
Translation from the French language edition: "Le marathon du messager - Histoire des vaccins à ARN messager" by Jérôme Lemonnier et al., © EDP Sciences 2022. Published by EDP Sciences. All Rights Reserved.

The English translation of this book from its French original manuscript was done with the help of artificial intelligence (machine translation by the service provider DeepL.com). A subsequent human revision of the content was done by the author. The text has subsequently been revised further by a professional copy editor in order to refine the work stylistically.
The use of general descriptive names, registered names, trademarks, service marks, etc. in this publication does not imply, even in the absence of a specific statement, that such names are exempt from the relevant protective laws and regulations and therefore free for general use.
The publisher, the authors, and the editors are safe to assume that the advice and information in this book are believed to be true and accurate at the date of publication. Neither the publisher nor the authors or the editors give a warranty, expressed or implied, with respect to the material contained herein or for any errors or omissions that may have been made. The publisher remains neutral with regard to jurisdictional claims in published maps and institutional affiliations.

Disclaimer: The opinions expressed in this article are those of the author and not necessarily those of the publisher

This Springer imprint is published by the registered company Springer Nature Switzerland AG
The registered company address is: Gewerbestrasse 11, 6330 Cham, Switzerland

To François Jacob,

To François Gros.

A little more than 60 years ago, on May 13, 1961, François Jacob and François Gros, researchers at the Institut Pasteur in Paris, were invited by two American laboratories to validate their hypothesis, and simultaneously reported the discovery of messenger RNA in the same issue of the prestigious journal Nature...

Where danger grows, that which saves also grows.

Friedrich Hölderlin
German writer, poet and essayist (1770, in Lauffen-am-Neckar—1843, in Tübingen)

About the Authors

Jérôme Lemonnier, who is son of a veterinarian and a pharmacist-biologist, is responsible, in the Hauts-de-France region (North of France), for promoting innovation and economic development projects in the health and textile industries. A senior civil servant, he currently works at the crossroads between science and industry.

Nicolas Lemonnier Jérôme's cousin—whose mother is a public health doctor and the father a research professor in immunology at the Institut Pasteur in Paris—has been immersed in biology since his early childhood. While remaining in the scientific field, he nevertheless chose another path and became an engineer. He has co-founded two start-ups in the fields of consumer web, deeptech, and musictech.

Gilles Charrot is a cartoonist who has spent a good deal of his career working for the written press. Today, he takes part in congresses and other seminars, mainly medically oriented, during which he reacts immediately to the comments of the various speakers. In addition, he is also a French editorial secretary, which means that he is in charge of correcting spelling and syntax, tracking down typos, taking care of typography, and reformulating, if the need arises.

Acknowledgements

The authors warmly thank Steve Pascolo and Chantal Pichon, both for the time they kindly granted them during the exchanges prior to writing the manuscript and for the answers they gave to numerous questions. Their expertise on the subject, their scientific objectivity, and their availability were extremely valuable.

The authors also thank Jean Bénard, biologist and researcher in molecular pathology (Gustave-Roussy cancer center, Villejuif, France), who believed in this family project and gave it his unfailing support. He also provided a first scientific proofreading—attentive and methodical—of the book, with an always benevolent eye.

Of course, the book would not be what it is without the touch (and, sometimes, the claws that go with it!) of cartoonist Gilles Charrot, who provided a dose of affectionate impertinence. In addition to his inimitable pencil stroke, our talented illustrator is also, and still professionally, an editorial secretary; thus, throughout the reworkings of the manuscript, he provided this other know-how without counting his hours, nor ever losing his good mood during this adventure: a big thank you to him.

Jérôme Lemonnier would also like to thank his wife Aurélie, who has been patient, attentive, and supportive throughout this beautiful undertaking; she was the cornerstone of this work.

Nicolas Lemonnier, for his part, thanks his parents—François Lemonnier, a former research professor in immunology at the Institut Pasteur in Paris, and Jacqueline Lemonnier, a former public health inspector—who gave him his curiosity and a taste for understanding living things.

Contents

Foreword

This book traces the decades of research and innovation that have led to the development of an entirely new technology—mRNA vaccines—effective against Sars-CoV-2. These vaccines are, today, on the way to stopping the COVID-19 pandemic.

Jérôme and Nicolas Lemonnier, respectively, an actor in the promotion of scientific innovation and an entrepreneur, have investigated the history of this therapeutic innovation, a major one for humanity, with the French pioneers Steve Pascolo and Chantal Pichon. They meticulously and thoroughly retrace each of the contributions that have marked this astonishing "marathon of the messenger" since the discovery of mRNA by François Gros, François Jacob, and Jacques Monod in 1961. For all of these pioneers, imposing this new technology on their incredulous scientific community was a real test. The role of the biotechs and their financiers, who took great risks, and the results obtained were decisive in overcoming the ambient skepticism.

The paths of innovation are notoriously sinuous and sometimes disjointed, but the road to developing mRNA vaccines has been a veritable maze. The objectivity of the documentary sources highlights the true complementarity between the research and discoveries on both sides of the Atlantic. Through the history of mRNA vaccines, the two authors offer us a balanced analytical and critical vision of scientific progress in this era.

Scientists and policymakers alike will appreciate this book, which will hopefully bring new ideas to governments (and various research and innovation ministries) around the world, as they learn from the COVID-19 crisis and implement original fast-track pathways designed to deliver the new technologies essential to protecting and preserving our planet's living world, in all its forms!

Faro, Portugal

Pierre Meulien
Former Executive Director
of Innovative Medicines Initiative
(IMI)

Preface

Revealed knowledge has no other object than obedience,
and is thus entirely distinct from natural knowledge, both
in its object and in its principles and means.

Baruch Spinoza (1632–1677)

The Marathon of the Messenger is an analytical, critical, and objective approach to the history of messenger RNA vaccines. The discussion and the explanation are based on a wealth of available documentary sources, namely patents and medical articles that can easily be found on the Internet. The objective character of this analysis seems to us to be the best response to any suspicion of bias or arbitrariness.

If the book tells a story, it is not, in fact, based on any kind of *storytelling* likely to distort reality. It is not a fabulous and bewitching story, intended to capture the attention, nor even to carry away the credulous adhesion, of the reader. We are a long way from any sort of "mythological" story, which, for example, would provide us with a hero, a "savior of the world," who, misunderstood by all, fights against adversity and emerges triumphant. The emotional springs of such narratives constitute, in our eyes, a mystification, whose final effect is to prevent the distance necessary to engage with the thought, and to deprive the citizen of the time necessary for critical reflection.

We contest the substance of such accounts, which are, above all, based on the subjectivity of the characters. To echo them is to risk manipulation, even disinformation, sometimes without being fully aware of it.

We believe that such accounts do not provide a coherent, methodical, and well-founded vision of the adventure of messenger RNA vaccines. In the history of science, as in political or social history, a rigorous and objective analysis must prevail. As for historical criticism, it must also be applied to the history of scientific discoveries.

For this reason, we have chosen not to resort to a multiplicity of interviews with scientists and researchers. This avoided a "cataloged" and subjective account, in which each one could have tended to assert that he alone was worthy of esteem and recognition. The book is not, therefore, journalistic in nature.

On the contrary, we sought to acquire a deep understanding of our object of study. To define it well was, for us, the primordial condition for elucidating its history, understood here as a collective human adventure. This is how the real complementarity between the research and our individual discoveries appeared to us.

Finally, the development of messenger RNA vaccines is a formidable demonstration that, in everyday life, science, industry, economy, and culture are fields that, far from being separate, are actually closely linked. Scientific and economic progress and freedom of thought and speech are therefore inseparable.

The Marathon of the Messenger also aims to raise awareness of this fact.

The authors wish to consider themselves as acting in accordance with the philosophy of Baruch Spinoza, who also believed that humanity should use reason to understand the things of this world, and that, correspondingly, we should leave aside superstition, which is a source of prejudice and intolerance. It is certain, in our opinion, that, in today's world, humanity is more easily governed by superstition; nevertheless, we hope that, through engaging with this book, readers will make good use of their reason to come to a solid understanding of the progress made possible by the use of messenger RNA for therapeutic purposes.

In this sense, our book is also part of the fight against prejudice and contempt for science.

Wasquehal, France   Jérôme Lemonnier

Introduction: The Marathon of the Messenger RNA...

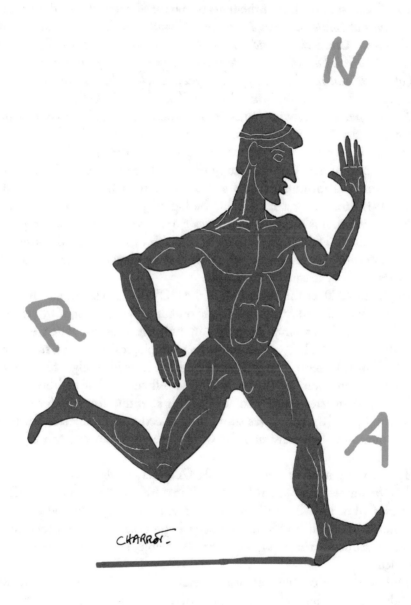

In his *Moral Works*, Plutarch reports the following historical fact about the messenger Eucles, who returns out of breath from the battlefield of Marathon, in 490 B.C., to inform the Athenians of the Greek victory over the Persians: "*Heraclides of Pontus says that Thersippus of Ereus brought the news of the battle of Marathon. Others claim, and it is the greatest number, that it was Eucles. They say that he arrived at Athens still steaming with the blood of the enemies; that he fell of tiredness at the door of the magistrates, to whom he said only these words "Rejoice, we have overcome", and that he fell dead at their feet*".

The runner Eucles, who transmits the good news, and thanks to whom the whole city knows the Greek victory against the Persian "barbarians," resembles, in a certain way, the messenger RNA (mRNA).

In the case of Sars-CoV-2, a copy of part of the messenger RNA of this viral intruder is formulated into a vaccine. It then helps humans to oust the COVID following a remarkable biological process, which involves the immune system. It is a journey on the scale of the infinitely small, comparable to that of Eucles, which separated Marathon from Athens. And considering the decades-long development of mRNA technology for vaccination, the historical allegory is equally applicable.

In March 2020, at the beginning of the COVID-19 pandemic, who could seriously have believed in a prophylactic outcome brought about by synthetic mRNA vaccines? Vaccines promised in only six months, when everyone thought it would take years, even with technological tools far superior to those of Louis Pasteur? The greatest skepticism reigned. Only a few initiated researchers were aware of the technological difficulties that had been solved over the past quarter century to achieve this scientific feat: the formulation of a synthetic mRNA that was sufficiently effective to be injected for therapeutic purposes. This illustrates the work, ingenuity, and strength of will of the human being.

The development of synthetic mRNA vaccines also demonstrates the collective dimension of scientific work. It involved researchers from different countries who, little by little, relied on methods recognized and shared by all: the formulation and verification of hypotheses through experimentation and deductive reasoning based on observable data, to advance objective knowledge.

This work was carried out over several decades. First of all, there were the adventurers, who had the immense merit of clearing a *terra incognita* for biology, at a time when the adventure could be frightening or appear useless or too risky.

These researchers were followed by physicians who had the tremendous desire to use mRNA to offer new therapies and hope to patients. Their support is now leading to more personalized, more refined, and more targeted therapeutic approaches.

All have contributed to the progress of biology and medicine. All of them also help human beings today to better understand the most fundamental biological mechanisms, which guarantee, in all living beings, the activities essential to life.

These biologists and physicians have made COVID mRNA vaccines a major achievement, which has already had a significant impact on health and economic policies in all countries of the world. At a time when the COVID-19 pandemic is shaking up the world and panicking governments, these innovative and effective vaccines, which were the first to appear, are the most effective treatments; all countries are now clamoring for them, as they appear to be essential for achieving collective immunity and boosting the economy.

This book aims to tell the story of vaccines based on synthetic mRNA, a story that has been little known until now. It aims to make known its founders and their difficulties in imposing this new therapeutic approach. It recounts this saga, in which the heuristic genius of humanity demonstrates its power to find solutions to questions, enigmas, headwinds, and uncertainties that arise at the limits of the known world. This book is an opportunity to show gratitude and appreciation to these pioneers, both for the work already done and for the work made possible for the future.

These scientists were, as General de Gaulle said of Winston Churchill, "*the great champions of a great enterprise, and the great artists of a great story.*"

Part I

Messenger RNA, An Essential Role and Challenge

1

A Short History of Vaccination

Vaccination consists of stimulating the body's immune defense mechanisms against a micro-organism capable of causing disease.

The understanding of the biological mechanisms involved in vaccination has been slow and progressive.

Initially, the methods were empirical. It seems that, as early as the Middle Ages, variolization was practiced in China: this consisted of inoculating the contents of a vesicle from a patient with a weak smallpox infection into the person to be immunized. The result was uncertain: the mortality rate could reach 1–2%. The process was very controversial and was gradually abandoned in the eighteenth century.

The empirical method of vaccination developed by the English physician Edward Jenner in 1796 proved to be much more effective. Jenner had found that English farm women were often exposed to a mild ruminant disease called cow pox (*variola vaccina*), which was characterized by pustular spots on the cows' udders. The women farmers themselves had pustules on their hands, because cross-contamination occurred during milking. However, these women never got sick with smallpox. Edward Jenner came up with the idea of inoculating other people with the contents of the vesicles, which also made them immune to smallpox. This process spread rapidly, first in England and then in other countries. The word "vaccination" comes from this inoculation of vaccinia. The method is inductive: first start with the facts, then build a logical reasoning.

© The Author(s), under exclusive license to Springer Nature Switzerland AG 2023
J. Lemonnier and N. Lemonnier, *The Marathon of the Messenger*,
https://doi.org/10.1007/978-3-031-39300-6_1

At the end of the XIXth century, Louis Pasteur used the same inductive and experimental approach in his research on rabies. A chemist by training, he was the first to demonstrate the existence of disease-causing microorganisms by studying chicken cholera. He was the first to understand the principle of vaccination, based on the attenuation of the virulence of germs and the acquisition of immunity, first in chickens, then in sheep affected by anthrax, and finally in humans.

The stages of this great "leap" deserve to be briefly recalled. While working on rabies, Pasteur discovered the processes that attenuated the virus present in the brains of rabid dogs—for this, he needed a whole animal house of dogs... The rest is known: in 1885, Joseph Meister, a child from the East of France (the Alsace region) who had been bitten by a rabid dog, was brought to him. Pasteur gave him several injections, first of very weakened viruses, then of more and more virulent ones. It was a real success. A larger-scale experiment was carried out; as the great oncologist Maurice Tubiana (1920–2013) writes in his *History of Medical Thought*: "*It was a triumph, despite the outcry of doctors who raged against Pasteur, who was not a doctor and who dared to inject his vaccine*" [1].

Pasteur's rabies vaccine is therefore a live attenuated vaccine: the patient is injected with a modified, harmless version of the pathogen, which then stimulates the patient's immunity, according to the immune response mechanisms described in the first part (see *below*, section I, Chap. 5, "The Immune Response"). This type of vaccine was developed during the twentieth century: BCG (1921), polio vaccine (1952), etc.

There are also vaccines that are made from killed or inactivated viruses, but that still retain immunogenicity, such as the pertussis vaccine developed in 1926. These vaccines generally contain more adjuvants, and boosters are more often required. For Sars-CoV-2, the Chinese vaccine Sinovac is based on this approach.

Among all these vaccines, MMR (measles/mumps/rubella), developed in the 1960s, deserves special mention, as it was the first mRNA vaccine in the world. As the measles virus is an RNA virus, the corresponding vaccine (mandatory in France) is also an RNA vaccine, like the Sars-CoV-2 vaccine. However, in the case of measles, it is a natural RNA, whereas it is a synthetic RNA in the case of Sars-CoV-2. As Steve Pascolo, an immunologist and researcher at the University Hospital of Zürich, whose role in the history of vaccines will be discussed in detail later, explains: "*The MMR vaccine, measles, mumps, rubella, works with attenuated RNA viruses. When injected, the viruses deliver their messenger RNA into your cells. (…) Produced in fertilized eggs, MMR-type vaccines contain a lot of RNA, lipids, and various proteins.*

In contrast, the new RNA vaccines contain only the purest RNA molecule, and four lipids. In other words, the messenger RNA version [used against Covid-19] is much purer and safer than the naturally produced RNA vaccines (…)" [2].

Currently, three other types of vaccines have been developed that compete with mRNA vaccines for the treatment of Covid-19:

- DNA vaccines, characterized by a modification of DNA by genetic engineering to induce an immune response following the production of a specific antigen. For the moment, DNA vaccines are still in the experimental stage;
- viral vector vaccines, the production of which consists in integrating the gene for a viral protein into the genetic makeup of another virus. This principle is the basis of the Astrazeneca vaccine against Sars-Cov-2: the gene for the spike protein[1] is introduced into the genome of a chimpanzee adenovirus genetically modified to limit its replication. The Johnson & Johnson vaccine, based on the same principle, uses a genetically modified cold adenovirus to limit its replication. Thus, the vaccine makes it theoretically possible to integrate a part of viral RNA into the human genome, via a DNA virus responsible for a disease that has never before affected humans. This manipulation has absolutely nothing natural about it and is, in fact, genetic engineering. Paradoxically, it did not raise any scientific or ethical questions in France, with the notable exception of the late Professor Axel Kahn, who, as early as November 2020, insisted that Astrazeneca, Johnson & Johnson and Sputnik vaccines were "GMO vaccines" [3]! In this case, such a process could, in the long term, have an impact on the proper functioning of our immune systems. Very rare, but serious, thromboses linked precisely to a dysfunction of the immune systems of certain patients have already been identified as a side effect of the Astrazeneca and Johnson & Johnson vaccines. This risk is not incurred with mRNA vaccines;
- protein-based vaccines, based on the modification of viral proteins. In the case of Sars-Cov-2, the Novavax vaccine is based on this principle: the modified spike protein is integrated into nanoparticles, which then trigger the body's immune response.

Thus, mRNA vaccines represent the culmination of progress in vaccination over the past 200 years.

---

[1] The viral spike protein is a large transmembrane protein containing 1273 amino acids. It has played an important role in the development of anti-Covid mRNA vaccines: see *below* Part III, Chap. 19, "2020: The Triumph of Anti-Covid mRNA Vaccines".

They reflect an increasingly sophisticated understanding of how cellular mechanisms work, leading to targeted and clean preparation of the immune system. As we will see, this vaccine format results in the expression of purified proteins. Moreover, this type of vaccine requires fewer adjuvants and fewer animal experiments.

To understand what these vaccines are, let's first look at what mRNA is, explaining how it differs from the DNA in every living being.

References

1 Maurice Tubiana, *Histoire de la pensée médicale. Les chemins d'Esculape*, Éd. Champs Flammarion, 1997, p. 220.
2 Steve Pascolo, interview on French radio *France Inter,* January 4, 2021, available on the Internet at https://www.franceinter.fr/societe/vaccins-utiliser-l-arn-messager-n-est-pas-nouveau-c-est-savoir-le-fabriquer-qui-est-nouveau.
3 Axel Kahn, «Les vaccins d'AstraZeneca, Johnson & Johnson et Spoutnik sont des vaccins OGM», in *L'Usine Nouvelle*, article of November 27, 2020, available at: https://www.usinenouvelle.com/article/les-vaccins-d-astrazeneca-johnson-johnson-et-spoutnik-sont-des-vaccins-ogm-explique-axel-kahn.N1033634.

2

From DNA to RNA

The paths to scientific discovery and innovation are notoriously complex and winding, and those that led to the discovery of DNA and messenger RNA (MRNA) are no exception.

Let us recall, then, the most eminent facts that led to the fundamental principle of molecular biology.

In 1869, a Swiss doctor, Friedrich Miescher, who was almost deaf and, partly for this reason, unwilling to pursue a career in health care, while in the kitchen of the castle of Tübingen (Germany), which had been transformed into a laboratory, discovered a substance rich in phosphate and nitrogen from the nuclei of blood cells obtained from patients treated at the Tübingen hospital. He claimed that this substance, which he called "nuclein", was neither a sugar, nor a lipid, nor a protein: it was a new biological substance. Back in Switzerland, after two years of research in Tübingen, he continued his research in Basel, this time using cells of the sperm of the Rhine salmon. He isolated nuclein again. His students and successors continued to study nuclein, to define its composition and content: it consists of four basic elements called adenine, thymine, guanine and cytosine, which are associated with a phosphate bond.

In 1953, James Watson and Francis Crick, from the University of Cambridge (Great Britain), discovered the structure of DNA: an antiparallel double helix characterized by the pairing of the nitrogenous bases cytosine—guanine and adenine—thymine. **The particular structure of this double helix is described in the box "The skeleton of DNA and RNA".**

© The Author(s), under exclusive license to Springer Nature
Switzerland AG 2023
J. Lemonnier and N. Lemonnier, *The Marathon of the Messenger*,
https://doi.org/10.1007/978-3-031-39300-6_2

They received the Nobel Prize in Medicine in 1962 *"for their discoveries concerning the molecular structure of nucleic acids and their importance for the transfer of genetic information in living material"* [1].

The stages of scientific research that led to the discovery of mRNA are well highlighted by Matthew Cobb in his article *"Who discovered messenger RNA?"* If, since the end of the 1940s, many research teams contributed to the discovery of messenger RNA and to the demonstration of its role in the transmission of genetic information from DNA to protein [2], we often only remember the two articles published simultaneously in the journal *Nature* on May 13, 1961, which related this discovery [3, 4].

Just after these two articles, in June 1961, François Jacob and Jacques Monod published another article in the *Journal of Molecular Biology* in which they explained the nature and functions of mRNA [5]. In particular, the two authors indicated that the protein regulatory system *"seems to operate directly at the level of the synthesis by the gene of a short-lived intermediate, or messenger, which becomes associated with the ribosomes where protein synthesis takes place."* The essential point here is the transfer of information: the meaning of the message transmitted is the key point. The mechanisms of gene expression and regulation, including messenger RNA (mRNA), which links the genetic information encoded in DNA to proteins, are thus well understood.

Three years later, in 1965, these two French researchers received the Nobel Prize in Medicine, along with André Lwoff, also from the Institut Pasteur in Paris, for having, according to François Jacob, *"filled the gap between the gene (unit of heredity) and the character"* and understood *"by what mechanism the synthesis and behavior of proteins, which are the essential constituents of living material, were remotely directed"*.

Discoverers of mRNA:
Jacques Monod, François Jacob,
François Gros and André Lwoff.

More precisely, let's insist here on the fundamental role played by the gene, in order to understand how our DNA can be expressed and translated into each protein produced by our organism! The gene represents the functional unit of genetic information. Thousands of genes are found on each of our forty-six chromosomes; they constitute our genetic identity. For each cell of our organism, the same gene may or may not be expressed. For example, for a pancreas cell, the gene that codes for insulin production will be expressed. This "insulin production" gene will, of course, be present in the DNA of other cells in the body, but it will not be expressed in those cases. So, genetic

Fig. 2.1 François Gros around year 1960. Copyright Institut Pasteur reproduced with permission

expression is based on the production of the corresponding messenger RNA, *"which makes the link between the gene and the character"*, according to the luminous formula of François Jacob. Each human cell contains hundreds of thousands of messenger RNA molecules.

This specific RNA, called "messenger", acts as an intermediary between DNA and protein synthesis: it is the photocopy of the page of a book, which is destroyed after use. The genetic information thus passes from DNA to messenger RNA and from messenger RNA to proteins (Figs. 2.1 and 2.2).

Fig. 2.2 François Jacob. Copyright Institut Pasteur reproduced with permission

2.1   The Skeleton of DNA and RNA

In eukaryotic living beings (animals, plants, fungi and protozoa), DNA is located mainly in the chromosomes in the nucleus of each cell.

The nucleus, an "organ" that contains the information necessary for the functioning of any cell, contains the 23 pairs of chromosomes (in humans) that represent the genetic heritage of each individual. It determines the anatomical, physiological and, in part, behavioral characteristics of an individual.

All this information is stored in very long molecules similar to a very long necklace of several million pearls of four different colors: deoxyribonucleic acid, or DNA. The characteristic structure of DNA, organized in a double helix known as "Watson–Crick", includes four nitrogenous bases that pair with each other one strand at a time: adenine, thymine, guanine and cytosine, A, T, G, and C. It is this particular sequence, unique to each individual, that actually defines its own biological identity.

Each strand is structured in a linear but flexible skeleton, made up of the repetition of an elementary motif: phosphate-sugar-nitrogenous base. The sugar is deoxyribose for DNA and ribose for RNA. The deoxyribose contains one less oxygen atom than the ribose, hence its name.[1] The complementarity of the two strands is the guarantee of the maintenance of the genetic information: the sequence of a strand that disappears will be copied from the preserved complementary strand and will restore the lost bi-helical structure.

Unlike DNA, mRNA is single-stranded; it is a copy of the single coding strand of DNA, but it is written in a new syntax: as a sugar, ribose replaces deoxyribose, and, among its four nitrogenous bases, uracil replaces thymine. This molecule, which combines A, U, G, and C, is the intermediate and transient template that will be read by the translational machinery to synthesize proteins.

References

1 Quote from the following website: "The Nobel Prize in Physiology or Medicine 1962" [archive], at www.nobelprize.org, 1962 (accessed March 29, 2015): "*The Nobel Prize in Physiology or Medicine 1962 was awarded jointly to Francis Harry Compton Crick, James Dewey Watson and Maurice Hugh Frederick Wilkins 'for their discoveries concerning the molecular structure of nucleic acids and its significance for information transfer in living material'*.

2 Matthew Cobb, *"Who discovered messenger RNA?"*, Current Biology, Volume 25, Issue 13, 29 June 2015, Pages R526-R532. DOI: https://doi.org/https://doi.org/10.1016/j.cub.2015.05.032.

3 François Gros, H. Hiatt., W. Gilbert, Kurland., R.W. Risebrough, & James D. Watson, "*Unstable Ribonucleic acid Revealed by Pulse Labelling of Escherichia coli,*" Nature, 13 May 1961, vol. 190, p. 581–585. DOI: https://doi.org/10.1038/190581a0.

4 S. Brenner, François Jacob & M. Meselson "*An Unstable Intermediate Carrying Information from Genes to Ribosomes for Protein Synthesis*", Nature, 13 May 1961, vol. 190, p. 576–581. DOI: https://doi.org/10.1038/190576a0.

5 F. Jacob, J. Monod, "*Mechanisms of genetic regulation in protein synthesis*", in Journal of Molecular Biology, vol. 3. June 1961, Pages 318–356. DOI: https://doi.org/10.1016/S0022-2836(61)80072-7.

---

[1] **Nucleoside** is the element consisting of a nitrogenous base associated with a sugar.

3

Messenger RNA (mRNA): From Transcription to Protein Translation

The mRNA ensures the expression of the genetic information specific to each individual. This expression of information is carried out during two essential steps, transcription and translation.

i. *The transcription*

Transcription is the synthesis of RNA from DNA, carried out in the nucleus of each cell by specific enzymes called RNA polymerases. These enzymes polymerize RNA molecules from the DNA sequence using a specific code in which one of the four constituent bases of DNA (A, T, G and C), thymine (T), is replaced by uracil (U). RNA polymerases are the "photocopiers" of the big "book" that is DNA. Thus, the sequence of information specific to a gene, and to its function in the organ, is copied into an mRNA molecule. This transient message, once used by the machinery to make the desired protein, will be destroyed. It is a very fleeting messenger, which fulfills its mission in a brief moment!

This mRNA, which is written from left to right as 5′ and 3′, respectively, includes different parts:

• at the ends, two parts that characterize the mRNA: the "5′ cap" (which includes a guanine molecule, G) and the "poly (A) tail" 3′ (which includes several tens of adenine molecules, A);

J. Lemonnier and N. Lemonnier, *The Marathon of the Messenger*, https://doi.org/10.1007/978-3-031-39300-6_3

- sequences not translated into proteins, called UTR (*UnTranslated Region*): the "5′ UTR" and the "3′ UTR". The latter sequence plays an important role in ensuring the stability or instability of the messenger RNA molecule, which is important for the development of a vaccine based on this molecule. Small non-coding RNA molecules, called microRNAs, specifically target this 3′ UTR sequence of the messenger RNA. In doing so, they play an important role in the degradation of the messenger RNA molecule, which is why the choice of this 3′ UTR sequence is so important for the manufacturers of anti-Covid vaccines. Its optimization can lead to more stable mRNAs, and thus more efficient mRNA vaccines;
- finally, the region that counts the coding segments for the translation of proteins.

To produce synthetic mRNA in vitro, we use the DNA present in plasmids, which, in bacteria, are circular and autonomous DNA molecules. This template is transcribed into RNA by RNA polymerase. At the same time, or afterwards, the characteristic parts of mRNAs are added: a 5′ cap and a 3′ terminal sequence of the RNA—the poly(A) tail.

ii. *The translation of proteins*

Let's see how this translation is done.

Once transcribed, mRNA exits the nucleus through pores in the nuclear membrane into the cytosol, where the machinery is ready to translate the mRNA into proteins.

Translation is the synthesis of a polypeptide, i.e., a particular alignment of amino acids from mRNA. This process uses coding to read the "roadmap" of protein synthesis. On the mRNA sequence, the translational machinery is orchestrated by the recognition of triplets of bases, called codons.

There are 20 amino acids that are incorporated into human proteins. Every protein is built from a specific sequence of these 20 amino acids, forming a molecule of varying length. Each protein also has a specific structure and shape that determines its function in the body.

Each of these 20 amino acids corresponds, as mentioned above, to a triplet of RNA bases called a **codon.**

A codon is thus formed of three nucleotides, e.g., the sequence of a codon could be: guanine—guanine—adenine. For example, the amino acid glycine corresponds to the GGA triplet as well as the GGG triplet, while arginine corresponds to the CGC triplet as well as the CGG triplet. Therefore, if an mRNA has a sequence of the following bases: GGACGCGGGCGG bases, it will encode a glycine-arginine-glycine-arginine protein. It is thanks to this universal coding that the mRNA will be translated into a protein.

It is the sequence of amino acids making up a protein (which can be compared to a string of pearls, but ones of 20 different colors) that will dictate the function of the protein: insulin, antibody, hemoglobin, etc.

The cell builds the protein from the instructions it "reads" from the genetic message that is the mRNA. For a medium-sized protein, translation of a eukaryotic mRNA molecule takes one to two minutes. The cell has surveillance mechanisms that prevent the translation of incorrectly matured mRNAs.

In fact, translation takes place at the level of intracellular particles called ribosomes, whose function is to match an mRNA molecule with another type of RNA, called transfer RNA (tRNA). Each tRNA is attached to a particular amino acid. As translation proceeds, the tRNAs are aligned so that the message carried by each codon of the mRNA is translated into a specific amino acid. As translation proceeds, the amino acid molecules are linked together to form a chain. This is how the synthesis of a polypeptide takes place.

This polypeptide will then undergo various modifications, thanks to which it will acquire its final form to become a fully functional protein. For example, the specific folding of each protein, which gives it its own characteristics and functions (e.g., hemoglobin, insulin, keratin, myosin, Sars-CoV-2 spike, etc.).

The protein is then transported to its final destination, either inside the cell (cytoplasmic protein) or to the cell membrane (membrane protein) or outside the cell (secreted protein). Once the protein is translated, the mRNA is then degraded in the cell's cytoplasm. In most cases, degradation begins with the removal of the poly(A) tail or cap.

Genetic diseases, to the extent that they are characterized by DNA mutation, will most often result in the formation of altered mRNA, which will result in errors in protein translation. Agents known to be mutagenic also interact with DNA, resulting in mutations that lead to the synthesis of dysfunctional proteins. Mutagenic agents include ultraviolet radiation, X-rays, certain pollutants and chemicals, etc. Hence, the interest in synthetic mRNA for producing functional and normal proteins that will have a therapeutic virtue (cf. *infra* part II, Chap. 14, "Experiments and Clinical Trials Carried Out in Other Therapeutic Fields").

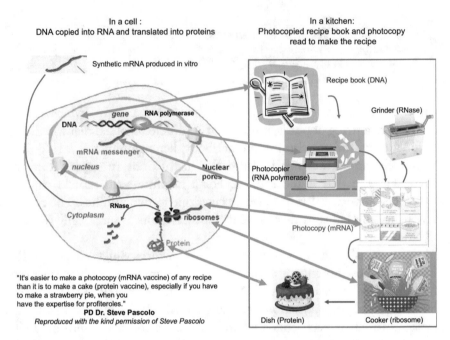

Fig. 3.1 Genetic expression and the making of a recipe (reproduced with Steve Pascolo's kind authorization)

Figure 3.1, which establishes an analogy between gene expression and the elaboration of a recipe, gives a complete and immediately understandable vision of all the molecular biological mechanisms involved.

4

The Mechanism of Viral Infection

Viruses cause many diseases. Some of them seem to have appeared suddenly in recent decades: see the box on "The AIDS virus and new viruses".

What are the processes involved in viral infection?

Although made up of genetic material (DNA or RNA), a virus has no life of its own: it cannot reproduce itself, in other words, replicate its genome. It is said to have a borrowed life. It can only develop inside a host cell of the organism it infects.

The size of viruses is very variable: from twenty nanometers for the smallest to about one thousand five hundred nanometers for the largest. Surrounded by a protein shell, some viruses also have a lipid envelope ("enveloped" viruses).

In the case of Sars-CoV-2, it is with a protein inserted into this external lipidic envelope of the virus, the spike protein, that it infects a given cell of the organism. Indeed, this protein binds to one of the proteins present on the surface of the rhinopharyngeal epithelial cells: the ACE2 receptor. Once this cell is infected, all its cellular machinery is put into service for the reproduction of the virus.

More specifically, in the case of an mRNA virus such as Sars-CoV-2, the virus also encodes a viral RNA replicase enzyme, which will duplicate this mRNA.

© The Author(s), under exclusive license to Springer Nature Switzerland AG 2023
J. Lemonnier and N. Lemonnier, *The Marathon of the Messenger,*
https://doi.org/10.1007/978-3-031-39300-6_4

Coronavirus family mRNA viruses, such as Sars-CoV-2, can sometimes mutate, even though their replicases are very faithful, because they have an error-checking mechanism. This accounts for the appearance of Covid-19 variants, as will be discussed in Part III (see Part III, Chap. 19 *below*).

At the same time, the viral mRNA will be translated into proteins of the capsid—the structure that envelops the viral genome—and into proteins of the virus's envelope itself (including the Sars-CoV-2 virus's famous spike protein, currently targeted by the synthetic mRNA vaccine).

The whole—formed by the duplicated viral mRNA and the translated viral proteins—then turns into a new virus, which will leave the cell by budding. The Sars-CoV-2 then infects other cells and, one by one, the process gets out of control.

"A virus can only grow inside a host cell."

4.1   The AIDS Virus and New Viruses

Some viruses, such as the human immunodeficiency virus (HIV), are retroviruses. HIV has an enzyme called retrotranscriptase, which converts viral RNA into DNA. This DNA can then be inserted into the DNA of the host cell. This mechanism is the basis of the HIV retrovirus infection. The inserted viral DNA remains permanently in the host cell's genome, where it is then transcribed into mRNA molecules that will be used for the next generation of retroviruses.

In addition, in the last few decades, new viruses, agents of infectious diseases, have appeared in the world. The rate of epidemics and pandemics has increased significantly for three main reasons:

- the propensity of existing viruses to mutate. These viruses may have a greater or lesser disposition to mutate. The mutation of a virus can, in particular, have the consequence of rendering ineffective the immunity previously acquired during the primary infection;
- intercontinental travel and migrations: populations are increasingly in contact with each other due to increased migrations. As a result, a virus that previously affected only an isolated part of a continent will have a propensity to spread and become a concern for the rest of the world;
- finally, the virus reservoir animals, deforestation, urbanization, meat consumption and industrialization are all factors that favor these epidemics. Thus, the proximity of animal farms (minks, ducks, chickens, etc.) to humans increases the risk of transmission of viruses from animals to humans. This is what happened with H1N1 in 2008, and probably with Covid-19.

Viruses that have emerged in recent years, in addition to HIV, include Ebola, Zika, Chikungunya, SARS (*Severe Acute Respiratory Syndrome*), MERS (*Middle East Respiratory Syndrome*), and, of course, Sars-Cov-2.

5

The Immune Response

The process of viral infection having been briefly described, it is important to present the mechanisms involved in the immune response of the infected organism.

Using the example of Sars-CoV-2, the viral mRNA, upon infection, is translated into viral proteins in the infected cell. These proteins produced by the cell, called antigens, trigger the immune response via specific cells called "antigen-presenting cells". This first step of the immune response is mainly carried out by dendritic cells, which play the most important role in antigen presentation. They have cytoplasmic extensions called dendrites, which provide an important surface for antigen presentation, hence their name. They are found in abundance in all organs, especially in the mucous membranes, lungs and intestines, liver, skin… Present in all tissues, dendritic cells are particularly abundant in lymphatic tissues.

In fact, the immune response elicited by an mRNA vaccine is based on the same defense mechanism of the body.

The human immune system is a complex and elaborate mechanism for the defense of the body against attacks by pathogens of all kinds (e.g., viruses and bacteria). The body's immune response is essential to ensuring the full effectiveness of mRNA-based vaccination. It can be compared to a country's defense system:

(i) the "Special forces" are the interferons, which intervene in very rapidly to seal off and neutralize an infected area;

© The Author(s), under exclusive license to Springer Nature Switzerland AG 2023
J. Lemonnier and N. Lemonnier, *The Marathon of the Messenger*,
https://doi.org/10.1007/978-3-031-39300-6_5

(ii) the "Gendarmerie" are the auxiliary T lymphocytes, which alert and coordinate the cells that will intervene in a more specific way (B lymphocytes and cytotoxic T lymphocytes);

(iii) the "Police" are the B lymphocytes, which intervene selectively to neutralize viruses and other pathogens;

(iv) the "Regular army" are the cytotoxic T lymphocytes, which use weapons of targeted destruction on infected cells;

(v) the "Intelligence service" is both the informers who keep an eye out for any dangerous signals that may appear in all the organs, and the members of the Police, Regular army and Gendarmerie who have recorded the previous attacks. These include macrophages, dendritic cells and B lymphocytes.

Immunologists therefore distinguish between non-specific immunity, known as "innate", and specific immunity, known as "adaptive".

- *Innate (or non-specific) immunity/inflammatory reaction ("Special forces" and "Intelligence service").*

The cytoplasm of human cells has "sentinels" that are highly reactive to the inappropriate presence of nucleic acids (as during an infection). These "sentinels" are intracellular receptors called PRRs (*Pattern Recognition Receptors*). They constitute a very fast intracellular defense line for the detection of pathogens.

In particular, these are receptors called *Toll-Like Receptors* (TLRs).

Toll-Like Receptors are transmembrane recognition proteins. They are located on the plasma membranes of immune cells or on the inner surface of various intracellular compartments, such as endosomes, which are vesicles that carry proteins into the cell and which do not contain, under normal conditions, RNA or DNA.

There are ten *Toll-Like Receptors* in humans. Discovered progressively during the 1990s, they are involved in the triggering of innate immunity and in the inflammatory reaction. The first demonstration of the role of TLRs in innate immunity was made in 1996, by Lemaître et al. [1], from the laboratory of Jules Hoffmann, who showed that Toll participates in antifungal immunity in Drosophila. Jules Hoffmann received the 2011 Nobel Prize in Medicine for this discovery.

The two TLRs involved in the recognition of exogenous single-stranded RNAs (such as Sars-Cov-2) are TLR 7 and TLR 8, both located on the inner surface of endosomes. TLR 3 is involved in the recognition of viral or parasitic double-stranded RNA.

In the presence of exogenous mRNA or DNA, these TLRs trigger an immediate (innate) and non-specific immune response (inflammation). Cellular activities, primarily protein production, are stopped in virtually all cells (except for certain specialized cells of the immune system). This triggers the activation of various cells, all of which are involved in innate immunity: mast cells, macrophages, dendritic cells, neutrophils, natural killer cells, monocytes, etc.

The main mechanism of action involved is phagocytosis; this is the cellular process by which cells can ingest solid particles of pathogens, with the aim of immediately destroying them. The most active cells in phagocytosis are macrophages, monocytes, neutrophils and dendritic cells. Phagocytosis also has a primary function: antigen capture, the first step in the body's adaptive response to infection.

Following this capture, dendritic cells undergo a maturation process characterized by the secretion of proteins that act as activating signals, the cytokines. **Their role is specifically described below in the box "The role of cytokines"**.

5.1   The Role of Cytokines

Cytokines are proteins of great diversity; their effect on the immune system is the subject of much discussion nowadays.

There are different types of cytokines; these include:

- chemokines. These cytokines provide certain chemical signals that induce the infiltration of white blood cells after a tissue injury (such as when fluid is injected into a muscle) or an infection. They tell the white blood cells where to go, where the injury occurred. In humans, there are about 40 different chemokines and more than a dozen chemokine receptors. In fact, it is certain that chemokines play an essential role in the circulation of white blood cells in the body and, consequently, in the proper functioning of our immune system: in fact, they are involved in regulating the circulation of lymphocytes in the lymph nodes;

- interferons. As stated in *Campbell Biology* (chapter "The Immune System"), "*these are proteins that provide an innate defense against viral infections. Cells infected with pathogens secrete these proteins, which induce neighboring cells that are not yet infected to produce substances that inhibit viral reproduction*" [2]. Interferons also block protein translation in infected cells, except for a certain type of specialized immune cell (see below, Part II, Chap. 12, "Experiments and Clinical Trials Against Infectious Diseases").

Their secretion appears to vary greatly from one individual to another in the case of mRNA vaccination. Depending on the formulation, dosage, and site of injection (subcutaneous or intravenous), the immune response will vary widely. It is certain that interferons play a role in T-cell activation; this is particularly true of type 1 interferon, the role of which is discussed *below* in section II, Chap. 12, "The Role of Type 1 Interferon in Triggering the Adaptive Immune Response". In this case, the impact of interferon secretion on antibody production, especially in the long term, is not known;

- interleukins, which are expressed by white blood cells and serve as messengers between immune system cells. The action of these cytokines has been the subject of debate among researchers for several decades, as it is not known precisely what role each of them plays in determining the immune response. However, there is evidence of an enhanced immune response with the use of interleukin 12, which justifies its current use in the experimental treatment of cancers and infectious diseases.

In conclusion, it appears difficult to classify all cytokines into categories, given their great diversity. However, it has been found that there are, in fact, links between the secretion of certain types of cytokines.

It should be added that the extraordinary diversity of the genetic heritages of individuals goes hand in hand with great differences in the immune responses induced by memory cells, which makes it very complex to draw global conclusions on the subject.

This is where the inflammatory response comes into play in the overall immune response, which is intended to facilitate the fight against the infectious agent and is usually manifested by a fever. This inflammatory reaction occurs in the case of localized infection, but also in the case of mechanical injury due to an intramuscular injection—as, for example, during vaccination against Sars-Cov-2—even if the injected material (e.g., modified mRNA) is not immunogenic. The secretion of the signaling molecules cytokines and histamine then ensures a more specific intervention of the immune system's guardians, the white blood cells. The dilation of blood vessels, which gives the characteristic redness of inflammation, actually facilitates their arrival at the site of the infection.

In this case, a more serious infection usually leads to fever, which can have harmful consequences for the body when it becomes too severe. When the immune system fails to overcome the infection, a very serious and potentially life-threatening inflammatory reaction occurs. This is known as septic shock and occasionally occurs in patients with severe Covid-19, among other cases.

This marks the beginning of a more specific intervention, specifically designed to target the offending infectious agent.

- **Acquired immunity and activation of cellular immunity ("Gendarmerie", "Police" and "Regular army")**

Thanks to acquired immunity, the organism defends itself in a specific way against a given pathogen (bacteria or virus). Three types of white blood cells (or leukocytes) are then involved and express themselves in activating cascades:

- T-helper lymphocytes;
- B lymphocytes;
- cytotoxic T lymphocytes.

Figure 5.1 describes the general functioning of the adaptive immune response.

- T-helper lymphocytes ("the Intelligence service" and "the Gendarmerie")

These lymphocytes become activated when they recognize antigens presented by dendritic cells or antigen fragments present in infected cells. In the case of TLR signaling, dendritic cells will migrate from the site where they encountered a pathogen to the draining lymph nodes, where they can activate T cells.

In detail, things happen as follows. In each cell, there is a sophisticated system for recognizing self and non-self called the major histocompatibility complex (MHC), for which its discoverers, including the Frenchman Jean Dausset, were awarded the Nobel Prize in Medicine in 1980. In case of infection, the MHC is altered, a fact that the helper T cells, in their vigilant surveillance, will not fail to detect. This recognition of the antigen and the alteration of the MHC, as well as the cytokines secreted by the infected cell, result in the activation of the T lymphocytes. The latter, in turn, activate either B lymphocytes (thanks to cytokines) or cytotoxic T lymphocytes (thanks to specific signals that they emit).

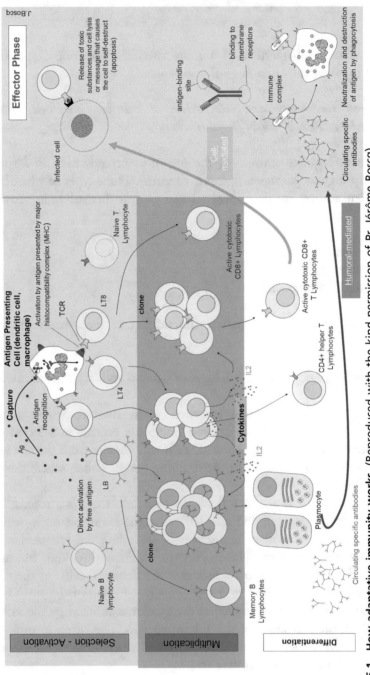

Fig. 5.1 How adaptive immunity works. (Reproduced with the kind permission of Pr. Jérôme Boscq)

- B lymphocytes ("the Intelligence service" and "the Police")

These cells directly recognize antigens from infectious agents, thus initiating the secretion of proteins called antibodies (or immunoglobulins). Each B cell contains approximately 100,000 antigen receptors; however, it should be noted that only one type of antigen receptor is found on a B cell. Therefore, the same type of B-cells, i.e., all with the same antigen receptor, are always involved in the fight against a given infection; other B-cells will be mobilized to fight another infection. Once activated, B cells proliferate and give rise to so-called "memory cells" (the "living memory" of our immune system, which keeps track of the viral infection) and to plasma cells, so-called "effector cells", which immediately begin to fight the antigen by producing antibodies.

It should be noted that, during certain viral infections, specific antibodies, known as "facilitating" antibodies, may promote the entry of the virus into certain cells of the immune system, in particular, the macrophages. This phenomenon of facilitating infection is described in the box "Facilitating antibodies".

- Cytotoxic T lymphocytes ("the Intelligence service" and "the army of the trade")

These cells detect infected cells via recognition of antigens and altered MHC, but in a slightly different way than helper T cells. When activated, they secrete proteins (perforins and granzymes) that "pulverize" the infected cell.

The introduction of a foreign mRNA into the cells of a living being can therefore trigger an immune response. In the case of synthetic mRNA vaccination, for example, lymphocytes will become active and fulfill their respective roles (producing antibodies, coordinating the immune response, and killing cells that express the targeted protein), and memory cells will keep track of this past reaction, which will help the organism more effectively to defend itself in case of infection.

"*The gap between the gene (unit of heredity) and the character*" having been filled, and the mechanism by which "*the synthesis and behavior of proteins were remotely directed*" having been understood, to quote François Jacob, let us now study the scientific and experimental pathway by which it became possible to use mRNA for therapeutic purposes. This pathway spans from the 1990s to the 2020s.

FRANÇOIS JACOB
FATHER OF mRNA

5.2   Facilitating Antibodies

According to the guide *Vaccins contre la Covid-19: questions et réponses*, intended for caregivers, and published by the Société de pathologie infectieuse de langue française [3], "In some infections, the presence of pre-existing immunity to the infection (of natural or vaccine origin) may favor severe forms of this infection: either because pre-existing antibodies facilitate the infection of immune cells (macrophages, in particular—the so-called "facilitating antibody" phenomenon), or because the direction of the immune response induced by the vaccine promotes a deleterious inflammatory reaction [...]".

The biological process by which facilitating antibodies operate is called opsonization: an antibody binds to an antigen of a pathogen; this facilitates phagocytosis by a macrophage (often a dendritic cell), and results in infection of the cell.

During its clinical trials of a vaccine against Covid, the German company BioNTech, which we will discuss later, probably considered the possibility of producing facilitating antibodies, which probably justified the conduct of several clinical trials. The following were tested:

(i) two "short" mRNA molecules encoding the only RBD domain (BNT162a1 and BNT162b1) (the RBD (*Receptor Binding Domain*) domain of the spike protein is responsible for the recognition and binding of the Sars-Cov-2 virus with the cell surface receptor ACE2);

(ii) a "long" mRNA molecule coding for the whole spike protein (BNT162b2);

(iii) a replicative RNA molecule (BNT162c2) (details on self-replicating mRNA are also given in section II, Chap. 12, "Experiments and Clinical Trials Against Infectious Diseases", *below*).

In the Phase-I clinical trial with healthy volunteers, only antibodies that neutralized the infection were found, not facilitating antibodies. BioNTech then continued its trials with the BNT162b2 vaccine. It can therefore be concluded that mRNA vaccines with an mRNA coding the spike protein do not induce the production of facilitating antibodies in the context of Sars-CoV-2.

References

1 Lemaitre, B., Nicolas, E., Michaut, L., Reichhart, J. & Hoffmann, J. "The dorsoventral regulatory gene cassette spätzle/Toll/cactus controls the potent anti-fungal response in Drosophila adults," *Cell* 86, 973–983 (1996). DOI: https://doi.org/10.1016/s0092-8674(00)80172-5.

2 *Campbell Biology*, 11th edition by Lisa Urry, Michael Cain, Steven Wasserman, Peter Minorsky, Jane Reece, Pearson Education, 2017.

3 *Covid-19 Vaccines: questions and answers* guide, from the Société de pathologie infectieuse de langue française, February 2021, pages 7–30. Guide is available at: https://www.infectiologie.com/UserFiles/File/groupe-prevention/covid-19/vaccins-covid-19-questions-et-reponses-spilf.pdf.

Part II

From Preliminary Studies to Clinical Trials

6

Promising Studies in the 1990s

In the years following the discovery of mRNA, the very idea of using it for therapeutic purposes seemed unthinkable within the scientific community. How could this be envisaged, when this molecule is so sensitive to RNases, RNA-destroying enzymes that are very abundant in the extracellular environment? And when its half-life in vivo is so short? For this reason, the opinion of scientists is unanimous: to introduce DNA that is stable by its nature into cells, yes, but RNA, no!

In 1990, Wolff et al. demonstrated for the first time that the intramuscular injection of "*naked*" nucleic acids (DNA and *mRNA*) into mice induced the expression, for a few days, of a protein encoded in muscle cells [1]. Robert W. Malone's role in the study must be emphasized here: he designed the experiments, assembled the plasmids, RNA and reagents, wrote the instructions and sent the samples to Dr Wolff at the University of Wisconsin, who then injected the products into mice with blinded samples. The precise role of Robert W. Malone is explained by Elie Dolgin in his article "*The tangled history of mRNA vaccines*", published in *Nature* on September 16, 2021 [2]. However, he is cited as the second author of the study, which appears to be an "exception" to the general principle regarding the importance of authors in scientific papers.

For scientific papers, the key individuals are the last author (usually the lab or group leader who came up with the idea for the project, secured the funding, and secured the infrastructure) and the first author (usually a student or post-doctoral researcher who carried out the project they were assigned

© The Author(s), under exclusive license to Springer Nature Switzerland AG 2023
J. Lemonnier and N. Lemonnier, *The Marathon of the Messenger*,
https://doi.org/10.1007/978-3-031-39300-6_6

and more or less contributed to its completion). The second and second-to-last authors may also have made a significant contribution. The other authors have often only participated in a distant way in the results presented in the article (punctual help with the experiments or contribution of some reagents useful to the progress of the project, for example).

However, after the Wolff et al. study, it was the injection of naked DNA that attracted the attention of the scientific community. It is from this perspective that the authors of the study mention for the first time the possibility that "*The intracellular expression of genes encoding antigens may provide alternative approaches to vaccine development*".

And in the years following the publication of this article, numerous works demonstrated the vaccine efficacy in mice of naked DNA injections. In contrast, the functionality of naked mRNA in these experiments was not noticed at first.

It was not until the French study conceived by Pierre Meulien and carried out by one of his collaborators, Frédéric Martinon (Inserm, Cochin Institute of Molecular Genetics, Pasteur Mérieux Sérums et Vaccins), in 1993, that the first article proposing and validating the use of mRNA in vaccines appeared. By injecting, by different routes (intravenous or subcutaneous), a synthetic mRNA coding for the "nucleoproteine" protein of the influenza virus, coated in lipid spheres, these researchers succeeded in inducing a cellular immune response (T lymphocytes) in mice against the virus responsible for influenza [3] (Figs. 6.1 and 6.2).

Thus, "*the present study is the first to demonstrate virus-specific induction of CTLs* [cytotoxic C lymphocytes] *by mRNA delivery* in vivo. *The results presented here highlight the utility of mRNA delivery systems for immunization with well-defined viral antigens. The interest of this new approach lies in the ease of preparation of liposomes containing mRNA*" (…).

This seminal work identified the fundamental methodological aspects that will determine the activity of researchers for nearly two decades: the site of delivery of the transcribed mRNA in vitro, the design of lipid nanoparticles that contain and protect the mRNA, and the lymphocyte response (after injection of the mRNA intravenously or subcutaneously).

The methodology proposed by Pierre Meulien and Frédéric Martinon offers a completely innovative solution. The injection of mRNA would help to avoid the production of protein antigens, an operation that involves long and costly steps (cell culture and purification); on the other hand, compared to a DNA vaccination (naked or in an attenuated virus), the process eliminates the risk of recombination within the individual's genome.

Fig. 6.1 Pierre Meulien

Fig. 6.2 Frédéric Martinon

The article was published in a review of modest impact, as the major scientific and medical journals were not interested in the study at the time. Curiously, the publication aroused only limited interest in scientists at the time as well, even though it was a simple process compared to the use of peptides or proteins carried by dendritic cells. Few scientists see this as a future avenue for the treatment of infectious diseases.

This breakthrough is not pursued by the authors themselves, who recognize a lack of reproducibility of this vaccination method. Indeed, the process did not seem to work in a robust way, nor did it work with antigens other than the influenza nucleoprotein. It was therefore abandoned.

Consideration of the use of mRNA to induce immune responses against cancerous tumors also began in the 1990s. The work of Martinon and Meulien was followed, in 1995, by a study conducted by Robert M. Conry and his colleagues from the American University of Birmingham (Alabama), and published in the journal *Cancer Research* [4]: the authors demonstrated an immune response after intramuscular injection into mice of recombinant mRNA encoding the carcinogenic embryonic antigen.

The carcinoembryonic antigen is a glycoprotein: it is a protein that includes carbohydrate motifs that are mainly found in cell membranes. It is involved in the process by which cells adhere to each other or to the surrounding environment. This antigen is a sign of a cancerous tumor when it is present in high levels in the blood.

Thus, the first evidence: an mRNA encoding a tumor antigen can help stimulate the production of antibodies specific to this antigen!

The potential of messenger RNA as an anticancer vaccine was also approached by a different methodology. In 1996, Eli Gilboa's team (Duke University, North Carolina, USA) used dendritic cells[1] as a vaccine vector. Purified from peripheral blood, white blood cell precursors of dendritic cells, called monocytes, were differentiated in vitro under specific conditions into dendritic cells.

Synthetic mRNA was transfected in three ways: by simple incubation of cells and naked mRNA, by liposomes (mRNA formulated with DOTAP liposomes), or by "electroporation" into these cells.

Transfection refers to the process of transferring genes, i.e., exogenous genetic material, into eukaryotic cells (especially animal cells), without using a virus as a vector. The electroporation process consists in subjecting the cells to a high voltage electric field that destabilizes the membranes and makes them permeable. The electric field also induces a movement of mRNA molecules

---

[1] As seen in part I, Chap. 5, dendritic cells are a particular type of antigen-presenting immune cell. They therefore induce the activation of the adaptive immune response.

that are negatively charged, which forces their entry into the cells. The mRNA transfected dendritic cells are then re-administered to the patient.

In this protocol, mRNA is only one of the compounds involved in the vaccine process; other authors have proposed peptides or proteins to "load" the dendritic cells. This highly technological therapeutic approach paves the way for research on the use of mRNA in oncology: developing vaccines that target tumor antigens.

References

1 Jon A. Wolff, Robert W. Malone, Phillip Williams, Wang Chong, Gyula Acsadi, Agnes Jani, Philip. L. Felgner, "*Direct Gene Transfer into Mouse Muscle in Vivo*," *Science,* March 23, 1990, vol. 247. DOI: https://doi.org/10.1126/science.169 0918.

2 Elie Dolgin, "*The tangled history of mRNA vaccines*," *Nature,* 16 September 2021, vol. 597. Springer ed.

3 F.Martinon, S. Krishnan, G. Lenzen, R. Magné, E. Gomard, J. G. Guillet, J. P. Lévy, P. Meulien, "*Induction of virus-specific cytotoxic T lymphocytes in vivo by liposome-entrapped mRNA*", *European Journal of Immunology,* July 23, 1993, pp. 1719–22. https://doi.org/10.1002/eji.1830230749.

4 Robert M. Conry, Albert F. LoBuglio, Marci Wright, Lucretia Sumerel, M. Joyce Pike, Feng Johanning, Ren Benjamin, Dan Lu and David T. Curiel, "*Character-ization of a Messenger RNA Polynucleotide Vaccine Vector*," *Cancer Research,* April 1995. DOI: Published April 1995.

7

The Use of mRNA—The Initial Technical Obstacles

The use of mRNA in vaccine applications has, in its early stages, posed significant problems, which were considered for some time to be difficult to overcome.

Indeed, mRNA has long been presumed an unstable molecule with a very short life span in the human body. These two characteristics alone made the design of therapeutic strategies based on its use unrealistic.

And yet, researchers have shown that:

- in vitro, mRNA, a single-stranded molecule, is very stable at room temperature in an environment strictly devoid of RNase enzymes (which destroy mRNA molecules in the extracellular environment): it can be kept for months at room temperature without significant degradation. Moreover, mRNA is the only biological molecule that can resist a temperature of 95° without losing its activity (DNA or proteins being denatured at this temperature). Under these conditions, its manipulation does not require expensive or complex equipment;
- the industrial ramp-up was achieved without major difficulties; this point will be studied more specifically in Part III below.

However, the use of mRNA for therapeutic purposes raised several important biological and biotechnological issues.

- Modalities of transport of the molecule in vivo towards the targeted cells.

© The Author(s), under exclusive license to Springer Nature Switzerland AG 2023
J. Lemonnier and N. Lemonnier, *The Marathon of the Messenger*,
https://doi.org/10.1007/978-3-031-39300-6_7

As mentioned earlier (see *supra* Part I, Chap. 5, "The Immune Response"), it is the antigen-presenting cells activating the helper T cells that trigger the immune response. Also, in order to reach the protein-producing ribosomal machinery, and to be recognized there as a natural mRNA, the synthetic mRNA must not undergo any degradation, neither by crossing the cell membrane, nor within its cytosol. This is why, since the 1990s, the use of lipidic nanoparticles called liposomes has been favored.

As early as 1989, the biotechnology company Vical, based in San Diego (USA), had succeeded in embedding messenger RNA in a lipid nanoparticle and introducing it into several types of cells: by this process, "*modified antiviral nucleoside analogues*" (DNA or RNA) "*can be integrated into the structure of liposomes, forming a more stable liposome complex that can deliver more drug to target cells with less toxicity*" [1]. These elements are also recalled in the article "*Cationic liposome-mediated RNA transfection*" by Malone et al. [2]. This article relates, in particular, certain investigations carried out at the Salk Institute at the end of the 1980s.

The information reported by Dolgin in the above-mentioned *Nature* article, "*The tangled history of mRNA vaccines*" [3], indicates that the instigator of the research was Robert W. Malone. Elie Dolgin reports, in particular, that "*at the end of 1987, Robert Malone carried out an important experiment. He mixed strands of messenger RNA with droplets of fat, to create a kind of molecular stew*". Whatever R. W. Malone's real involvement, it is unfortunate that he has come, thirty years later, to be so openly critical of messenger RNA vaccines, to which his research has nevertheless contributed.

In 1993, as part of their seminal work mentioned in the previous section, Pierre Meulien and his team also used liposomes to administer synthetic mRNA to mice: "*the mRNA encoding the nucleoprotein, obtained by* in vitro *transcription, was encapsulated by simple cholesterol/phosphatidylcholine/ phosphatidylserine liposomes by the detergent removal technique*" [4]. This "casing", which protects the mRNA during injection, guarantees its penetration into the targeted cells, then its effective release into the cell cytoplasm: the injected mRNA then reaches the ribosomal machinery to be translated into antigen in the presenting cell. This is how the immune reaction starts.

- Production of mRNA coding for the desired protein

The use of mRNA for therapeutic purposes means that the translation of a specific protein will be sought (cf. *supra* Part I, Chap. 3, "Messenger RNA: from transcription to protein translation"). This assumes that the mRNA fully expresses the genetic information it contains.

- Measurement of the effectiveness of the immune response

Treatment with mRNA involves optimizing the patient's immune response. In infectious diseases and cancer, an immune response is sought. In contrast, for gene therapies based on replacement proteins, the goal is to avoid the body's immune response. This response must be properly evaluated and, above all, regulated!

This issue has been the focus of research since the late 1990s. Thirty years of experimentation and controversy were required to assess the immunogenic impact of in vitro transcribed mRNA. As will be discussed in the following sections, CureVac and BioNTech have played a key role in this field.

- Optimization of the in vivo stability of the mRNA molecule

During the first experiments, first on mice and then on humans, the unstable nature of the "5' cap", "5' UTR" and "3' UTR" parts posed major difficulties for researchers. Various technical innovations, presented *below* (see Part II, Chap. 9, "Solutions for mRNA optimization"), have been made on the "5' cap" and the "3' UTR" region. They have contributed significantly to improving the intracellular stability of synthetic mRNAs.

- Mode of administration, and choice of injection site and dosage

When the use of mRNA began, the most relevant questions of the injection site and mode of administration (intramuscular, intravenous, intrathecal, intradermal, subcutaneous, lymphatic intranodules, etc.) were quickly raised. These factors may vary depending on the final destination of the mRNA: lymphatic system, nervous system, lung, liver, etc.

Following clinical trials, the intramuscular injection of mRNA into the deltoid muscle was chosen; this method of administration, which is totally accessible, is similar to that of the majority of vaccinations; it makes vaccine prophylaxis easy.

At the end of the 1990s, the really decisive breakthrough in the promotion of mRNA as a therapeutic tool for the treatment of infectious diseases and cancers was provided by a promising German start-up: CureVac.

References

1 Karl. Y. Hostetler, Raj Kumar, Louise M. Stuhmiller, *Lipid derivatives of anti-viral nucleosides, liposomal incorporation and method of use*, U.S. Patent (Vical. Inc), No. WO1990000555, international filing date: June 30, 1989, publication date: January 25, 1990. Available at: https://patentscope.wipo.int/search/fr/detail.jsf?docId=WO1990000555&tab=PCTBIBLIO&_cid=P20-KO9POE-50612-1.

2 R. W. Malone, P. L. Felgner and I. M. Verma, "*Cationic liposome-mediated RNA transfection,*" *Proceedings of the National Academy of Science of the USA*, August 1989; No. 86 (16): pp. 6077–6081. DOI: https://doi.org/10.1073/pnas.86.16.6077.

3 Elie Dolgin, "*The tangled history of mRNA vaccines,*" *Nature,* 16 September 2021, vol. 597. Springer ed.

4 F. Martinon, S. Krishnan, G. Lenzen, R. Magné, E. Gomard, J. G. Guillet, J. P. Lévy, P. Meulien, "*Induction of virus-specific cytotoxic T lymphocytes in vivo by liposome-entrapped mRNA,*" *European Journal of Immunology*, July 23, 1993, pp. 1719–22, DOI: https://doi.org/10.1002/eji.1830230749.

8

The Birth of CureVac: The Era of the Pioneers

The decisive impulse that led to the birth of CureVac was given in February 1996, by a professor of the Eberhard-Karl University in Tübingen. Located in the southern German state of Baden-Württemberg, it is a prestigious intellectual center, founded in 1477, and frequented at the end of the eighteenth century by the philosophers Hegel and Schelling, as well as by the poet and writer Hölderlin. Today, the university enjoys international recognition for the quality of its teaching in philosophy and natural sciences.

In the winter of 1996, one of the professors, Hans-Georg Rammensee, Chair of Immunology, proposed using direct injections of mRNA to vaccinate against cancers and infectious diseases, instead of modified cells or translated proteins. Inspired by the work of Eli Gilboa (see *Part II*, Chap. 6), Professor Rammensee nevertheless considered the transfection of mRNA into dendritic cells in vitro to be complex and difficult to implement: he envisaged injecting the mRNA directly into the body.

© The Author(s), under exclusive license to Springer Nature
Switzerland AG 2023
J. Lemonnier and N. Lemonnier, *The Marathon of the Messenger*,
https://doi.org/10.1007/978-3-031-39300-6_8

Since he needed liposomes for this purpose, he discussed his project with his colleague Günther Jung, professor of biochemistry in Tübingen. Both of them started to define the experimental methodology. Professor Jung was, at the time, the thesis supervisor of a young researcher in immunology, Ingmar Hoerr; thus, the professor entrusted his Ph.D. student with the study of the effects of the injection of mRNA included in liposomes on mice. Hoerr began his work in the second half of 1996. The first experiments on mice were started in collaboration with another PhD student, this one studying under the direction of Prof. Rammensee: Reinhard Obst.

In the spring of 1998, French postdoctoral researcher Steve Pascolo joined Professor Rammensee's research laboratory on a fixed-term contract and began work on the development of cancer vaccines.

In the late 1990s, Jung, Rammensee, Obst and Hoerr filed a patent to protect their discoveries, mainly the successful use of mRNA for vaccination. But when they filed their patent, the inventors discovered the existence of another patent that also concerned the vaccine use of mRNA formulated in liposomes—the one filed a few years earlier by Frédéric Martinon and Pierre Meulien. (Such a thing would be highly improbable today; however, one must put oneself in the context of a world in which the Internet had not yet disrupted access to communication, and, more particularly, communication within the scientific community, with publications only being accessible through articles printed on actual paper, and not the immense digital knowledge bases easily accessible today on Pubmed.)

BIRTH OF CUREVAC.

CHARROT.

Therefore, the four German scientists could not consider filing a patent for the use of liposomes for mRNA transfer. For this reason, the patent they finally filed at the World Intellectual Property Organization (WIPO) on March 14, 2001, only protected the use of proteins such as protamine for mRNA transfer into cells [1]. However, the patent would apply only to Europe; in this case, it was initially a handicap for CureVac not to have intellectual protection in the United States.

CureVac was founded in 2000. Ingmar Hoerr had the idea to create this start-up, whose business model is precisely based on the use of mRNA to heal (*cure*) and vaccinate (*vac*). At the time of its launch, CureVac had a professor from the University of Tübingen—Günther Jung -and three researchers- Ingmar Hoerr, Steve Pascolo and Florian Von der Mülbe; the latter was completing his thesis in Tübingen in Jung's laboratory. Under the recommendation of Steve Pascolo, Hans-Georg Rammensee joined the team a few months later.

CureVac's ambition was then to develop therapeutic treatments and vaccines based specifically on the use of mRNA.

This small biotech start-up was initially founded as a civil company under German law (*Gesellschaft bürgerlichen Rechts*, GbR), and not as a commercial company. It should be noted that German law stipulates that the partners of a GbR are personally liable, to an unlimited degree, for its activity, with their personal assets being at stake. The partners are jointly liable (*Gesamtschuldner*) for the obligations of the company towards third parties. The partners are all called upon to manage and represent the GbR. In short, CureVac was not originally registered in the German Commercial Register and had a character that could be described as "artisanal".

CureVac's growth was greatly aided in the early days by the *Jung Innovatoren* program. This was a competition organized by the federal state of Baden-Württemberg to encourage young entrepreneurs to start their own businesses. This type of program had already proved its worth in another German state, which was more advanced at the time in terms of economic support for biotechnology companies: Bavaria (Fig. 8.1).

CureVac's project was selected and the company received a grant from the federal state of Baden-Württemberg that provided two half salaries for researchers Ingmar Hoerr and Florian Von der Mülbe between 2000 and 2002; Steve Pascolo, for his part, was paid by a "Human Frontier" grant that he obtained in 1999, and then by another *grant*, i.e., a collaborative research program financed by the European Union on hemochromatosis coordinated by Professor François Lemonnier, in Paris.

Fig. 8.1 The founders of CureVac in 2002. From left to right: Hans-Georg Rammensee, Ingmar Hoerr, Steve Pascolo, Florian Von der Mülbe and Günther Jung

Thanks to this *Jung Innovatoren* grant, the fledgling company was also able to rent a laboratory free of charge from 2000 onwards, which consisted of a 50 m^2 room in the building housing the chemistry laboratories of the University of Tübingen. In 2001, CureVac bought its first machines and set up a "micro white room" of about 6 m^2. This was done in order to start production under GMP conditions (*Good Manufacturing Practices*), a European regulatory framework of principles to be followed by manufacturers in the production of medicines for human or veterinary use. The objective is to eliminate any risk of cross-contamination of products.

CureVac's scientific director, Steve Pascolo, then wrote the manuals of procedure that set out the methods of work in the start-up's laboratory. This was the first methodological guide for this previously non-existent mRNA technology. And under these conditions, the experiments proceeded as planned: the production of pharmaceutical mRNA in the small clean room did not pose any major difficulties.

However, vaccine development and initial mouse trials would continue to be conducted outside of CureVac for some time, in Professor Rammensee's laboratory.

Until 2003, CureVac was dependent on the Karl-Eberhard University in Tübingen for its operations. Afterwards, the biotech moved to brand new premises built by the state of Baden-Württemberg and the city of Tübingen, which was an important step in its development and the acquisition of its full autonomy.

In addition, between 2000 and 2002, CureVac decided to conduct experiments in conjunction with the Institut Pasteur in Paris and the University of Leiden in the Netherlands (*Universiteit Leyden*).

In fact, even though the results obtained by the inventors of CureVac by injecting mice with mRNA were promising, they still needed to compare them with those obtained after injecting DNA. Steve Pascolo therefore asked his former colleagues at Pasteur and one of his colleagues from Tübingen who had returned to the Netherlands to test his vaccine mRNAs. He himself would inject the mice with the mRNA while his colleagues injected the DNA.

The experiments carried out at the Institut Pasteur thus aimed to compare the immune responses induced by mRNA with those induced by DNA. Let's recall the principle of vaccination: to induce the activation and proliferation of lymphocytes specific to one or several antigens, hence the importance of comparing the magnitude of the adaptive immune response induced by DNA with that induced by mRNA.

The collaboration between CureVac and Institut Pasteur confirmed the induction of an immune response after injection of humanized[1] and wild-type mice with a formulation of mRNA coding for viral antigens (HIV and HBV). The results obtained, analyzed by Hüseyin Firat and Marie-Louise Michel, together with Steve Pascolo, clearly showed that mRNA injections induce an immune response, but that this response appears to be less strong than that with DNA injections. A joint CureVac—Pasteur patent was then filed at WIPO on January 20, 2003 [3]; however, the collaboration between CureVac and Pasteur researchers did not continue in the following years, and the application of this patent was not maintained.

The experiments carried out in 2000 at the University of Leyden were conducted by the Dutch researcher René Toes. The aim was to explore interest in the use of mRNA in the fight against HPV. The collaboration between CureVac and René Toes ended at the end of 2001.

---

[1] Humanized mice are mice that express a human gene; in the late 1990s, at the Institut Pasteur, "humanized" mouse lines were generated by Steve Pascolo and Professor François Lemonnier [2].

Also in 2001, CureVac indirectly obtained its first support from the European Union for the development of a pandemic vaccine. This support was granted within the framework of a research project giving rise to European funding (*grant*): the idea of such a grant is, indeed, to explore the feasibility of a new and improved technology, product, process or service. The beneficiary was officially Hans-Georg Rammensee's laboratory, where Steve Pascolo was working.

The *grant* in question—EU *grant* QLK2-CT-2001-01346—concerned the development of vaccines against swine fever caused by the CSFV (*Classical Swine Fever* Virus). This disease had been identified as a potential source of pandemic; it was especially responsible for very important economic losses in herds.

The *grant* was led by Prof. Matthias Buettner and aimed to assess the efficacy of different innovative vaccines by a comparative approach. Vaccines based on viral proteins on the one hand and nucleic acid (DNA and RNA) vaccines on the other hand were being developed and tested. In particular, the role of Hans-Georg Rammensee should be mentioned, as several formats of mRNA vaccines were developed at the University of Tübingen. However, the preferred technology remained that of Eli Gilboa, based on the transfection of dendritic cells with mRNA.

Barbara sings *

*The real song is:
"Göttingen".

The results were published in 2005 and are available at https://cordis.eur opa.eu/project/id/QLK2-CT-2001-01346/results .

The interest in using mRNA to elicit an immune response was reported. Pigs were vaccinated with mRNA and the results showed that they were effectively protected against CSFV.

In 2003, Hans-Georg Rammensee, Ingmar Hoerr, Florian von der Mülbe and Steve Pascolo continued their work on messenger RNA, as evidenced by patents filed at WIPO.

In 2006, after the major investment of 35 million by SAP's billionaire founder Dietmar Hopp through his company Dievini, Pascolo left CureVac to join the University Hospital of Zürich, where he optimized cancer chemotherapy protocols and continued his activities in regard to messenger RNA vaccines.

Just before Steve Pascolo's departure, the company achieved a world first: the validation of mRNA production under pharmaceutical conditions and in large quantities (Pascolo was the "*Kontrollleiter*" (which roughly translates from the German as "Oversight Manager") for the European Union). After the consolidation of the company, it succeeded in validating its methodological protocols and became a leading biotech at the European level.

The success of CureVac encouraged other research centers and biotechs to take an interest in mRNA vaccination. The challenge was to overcome the technical obstacles mentioned above (cf. *supra* Part II, Chap. 7, "The Use of mRNA: Initial Technical Obstacles"): during the period 2000–2020, all efforts were aimed at optimizing the mRNA molecule transcribed in vitro, improving its intracellular stability and optimizing its penetration into cells ("formulation") in order to *ultimately* increase the production of encoded proteins in the body. This optimization is totally conditioned by the therapeutic objective: is it vaccination or "replacement protein"?

References

1 Günther Jung, Ingmar Hoerr, Hans-Georg Rammensee, Dr. Reinhard Obst, *Transfer of mRNA using polycationic compounds*, German patent (CureVac), No. EP1083232, application date: September 9, 1999, publication date: March 14, 2001. Available at: https://patentscope.wipo.int/search/fr/detail.jsf?docId= EP13495749&_cid=P10-KM89OC-84545-1.

2 S. Pascolo, N. Bervas, J. M. Ure, A. G. Smith, F. Lemonnier, B. Pérarnau, "*HLA-A2.1-restricted education and cytolytic activity of CD8(+) T lymphocytes from beta2 microglobulin (beta2m) HLA-A2.1 monochain transgenic H-2Db beta2m*

double-knockout mice,", *The Journal of Experimental Medicine*, June 16, 1997; No. 185(12); pp. 2043–51. DOI: https://doi.org/10.1084/jem.185.12.2043.

3 Steve Pascolo, Ingmar Hoerr, Hans-Georg Rammensee, Florian Von der Mülbe, Marie-Louise Michel, Hüseyin Firat, François Lemonnier, *Immunogenic preparations and vaccines based on RNA*, Franco-German patent (CureVac, Institut Pasteur, INSERM and GENETHON), No. 022003059381, international filing date: January 20, 2003, publication date: July 24, 2003. Available at: https://patentscope.wipo.int/search/fr/detail.jsf;jsessionid=E862175F78F9C2D4A0DA4 F6CF48C4FA7.wapp1nA?docId=WO2003059381&tab=PCTBIBLIO.

9

Solutions for mRNA Optimization

It took a great deal of work before a synthetic mRNA could be administered to humans to protect them from or treat viral infections or cancers. With the help of competition, the three major biotechs CureVac, BioNTech and Moderna have contributed to the optimization of the mRNA molecule, from its design to its transport prior to injection. However, the decisive role of Drs. Ugur Sahin and Özlem Türeci of BioNTech must be highlighted.

Five technical barriers have been progressively lifted. They are presented *below* in a logical and non-chronological order.

- **1—Optimization of mRNA coding to produce the desired protein**

Coding sequence improvement is the first fundamental issue in the optimization of synthetic mRNA. Dr. Ugur Sahin and his BioNTech collaborators have been working on this critical topic for nearly two decades.

As early as 2006, Ugur Sahin noted that "*codon optimization is mainly based on the substitution of multiple rare codons by more frequent ones, which will encode the same amino acid. As a result, the rate and efficiency of the protein translation process are increased*" [1] (Figs. 9.1 and 9.2).

Why, then, does this optimization lead to greater efficiency in the protein translation process?

In concrete terms, mRNA optimization involves biological manipulations to promote the formation of codons containing guanine (G) and cytosine (C). This is based on the observation that the most abundant transfer

© The Author(s), under exclusive license to Springer Nature Switzerland AG 2023
J. Lemonnier and N. Lemonnier, *The Marathon of the Messenger*,
https://doi.org/10.1007/978-3-031-39300-6_9

Fig. 9.1 Özlem Türeci

Fig. 9.2 Ugur Sahin

RNAs (tRNAs) in humans (therefore, those favorable to obtaining the desired optimized mRNA) are complementary to the codons richest in guanine and cytosine. Codon optimization thus requires a decrease in the chemical element uracil (U), which can effectively induce less TLR stimulation and, consequently, a lesser inflammatory reaction.

This optimization is necessary, but has some limitations: indeed, it is closely dependent on the protein sequence to be expressed, and one should not remove too many highly immunogenic nucleotides from an mRNA vaccination perspective.

In recent years, CureVac and BioNTech have filed several validated patents on G- and C-enriched mRNAs (see, in particular, [2] for CureVac); the patents concerned would not have been validated if they had simply referred to an "mRNA with optimized codons", a notion that is too vague and does not necessarily appear innovative on first analysis. Moreover, the codons do not all have the same frequency: for example, in humans, the four possible codes for the amino acid Valine have the following frequency:

GTG—frequency: 0.47;
GTC—frequency: 0.24;
GTT—frequency: 0.18;
GTA—frequency: 0.11.

During the translation process, the more frequent the codon, the faster the transmission of genetic information from mRNA to protein. Thus, if the protein in question contains the amino acid valine, translation occurs more quickly if the corresponding codon is GTG. An mRNA designed with too frequent codons rich in G and C can eventually generate proteins that are translated too quickly and configured poorly; this prompts both caution when doing "codon optimization" and, possibly, the use of some rare codons to regulate the speed of protein translation. This fine balance in codon selection underscores that mRNA design is a critical step in producing a functional mRNA.

The optimization of the coding sequence can be obtained thanks to softwares that propose different sequences. From a given protein, companies (Twist Bio Science, Blue Heron, Gene Art…), will then submit a few sequences from a specific mRNA coding, to obtain the optimal quality transcripts. Certainly, internally, BioNTech, Moderna and CureVac have also developed the best processes for optimizing the coding sequences of their synthetic mRNA.

In conclusion, the biotechs CureVac, BioNTech and Moderna have each optimized similar mRNAs with slightly different sequences. These particular mRNAs are each protected by patents.

- **2—The improvement of mRNA molecule purification methods**

Optimization of the mRNA molecule also involves perfecting purification methods. As early as 2004, CureVac used the high performance liquid *chromatography* (HPLC) process, which separates mRNA bases, thus improving the quality of the mRNA for GMP production [3].

This mRNA HPLC purification process was patented by CureVac in 2007 [4].

It was then taken up by Katalin Kariko et al. in 2011 [5], with the objective of not triggering immune responses in the patient. The importance of the purification process is fully apparent here; it is stated in the abstract of the paper, "*although unmodified mRNAs were translated significantly better following purification, they still induced high levels of cytokine secretion. HPLC purified nucleoside-modified mRNA is a powerful vector for applications ranging from* ex vivo *stem cell generation to* in vivo *gene therapy*". While the distinction between modified and unmodified mRNA is discussed in the following section, it will also be seen, in Section II. Chap. 15, "Modified Versus Unmodified mRNA: Not Just a Scientific Issue", that this purification issue would play an important role in BioNTech's choice of a vaccine for Covid.

- **3—Better stability of the mRNA molecule**

From a biological point of view, the stability of an mRNA molecule ensures a good efficiency of the protein translation process and limits the destruction of the mRNA by RNase enzymes. From an industrial and medical point of view, the more stable the mRNA, the smaller the quantity of vaccine injected into the patient.

All regions of the mRNA can impact its stability. The coding region is very important: thus, poorly translated regions (with unfavorable codons) destabilize the mRNA. Stability is also improved when, during protein translation, there is a rapid uptake of mRNA by ribosomes. The "reading" of the mRNA then starts thanks to a loop that is made between the cap and the polyA tail. If this is not done properly, translation starts badly and there is a strong chance that the mRNA will be degraded.

At the same time, a given mRNA may be stable in one cell and very rapidly degraded in another. Not all the mechanisms that define the half-life of an mRNA in a given cell are currently known.

Thus, any and all regions of a given mRNA (the 5′ cap, the 5′ UTR, the region that contains the coding segments, the 3′ UTR and the polyA tail) can be enhanced to optimize the half-life of an mRNA in a given cell.

Some of the improvements that have occurred are particularly noteworthy.

- For the "5′-cap," while BioNTech made some interesting refinements in the early 2010s, the breakthrough innovation came from U.S. biotech company Trilink, which, in 2017, developed "CleanCap," a compound that optimally mimics the 5′-cap. This innovation led to the production, on an industrial scale, of a less expensive, more functional (from the perspective of protein stability and translation), and more effective mRNA for a number of therapeutic applications. This process also generates a five- to ten-fold increase in production during synthesis by transcription of the mRNA obtained in vitro.
- As to the "3′ UTR" part, between 2006 and 2015, BioNtech identified sequences in this part that confer greater stability of the transcribed mRNA and efficient translation into proteins [6, 7].

- **4—Significant progress in mRNA assay and mode of administration**

By improving the transport of the mRNA from the injection site to the cytoplasm of the cell, it has been possible to considerably reduce the doses injected. For example, for an mRNA vaccine against certain cancers tested in mice in 2000, a weak immune response was observed with 200 μg, whereas, in 2021, this response was significant in humans with only 30 μg of mRNA injected once (BioNTech anti-Covid-19 vaccination) [8].

This question is, in fact, inseparable from the mode of administration of mRNA, which has also improved significantly over the last two decades. Thus, the injection site and the route of administration of the mRNA vaccine have been optimized: intravenous, intramuscular, subcutaneous, etc. The objective is, once again, to optimize the release of mRNA into the cells according to the preferred routes and injection sites.

In a 2010 article in *Cancer Research*, Ugur Sahin and colleagues at BioNTech highlighted the value of injecting mRNA into the lymph nodes of mice, rather than intradermally or subcutaneously [9]. They found an immune response directed specifically against tumor antigens, both in helper T cells and cytotoxic T cells. Subsequent work by BioNTech would confirm the importance of the site and mode of administration for a better immune response in the patient.

One example is BioNTech's patent EP2830593, "RNA Formulation for Immunotherapy", filed for the treatment of cancerous tumors by immunotherapy [10].

This patent recalls the importance of the site of administration. It states: *"In order to provide sufficient uptake of the RNA by DCs, local administration of RNA to lymph nodes has proven to be successful. However, such local administration requires specific skills by the physician. Therefore, there is a need for RNA formulations which can be administered systemically, for example intravenously (i.v.), subcutaneously (s.c.), intradermally (i.d.) or by inhalation. From the literature, various approaches for systemic administration of nucleic acids are known. In non-viral gene transfer, cationic liposomes are used to induce DNA/RNA condensation and to facilitate cellular uptake"*. And further on, on the effectiveness of an injection into the spleen: *"For RNA based immunotherapy, lung or liver targeting can be detrimental, because of the risk of an immune response against these organs. Therefore, for such therapy, a formulation with high selectivity only for the DCs, such as in the spleen, is required"*. Injection of the mRNA into the spleen therefore results in a much greater immune response than injection into the lungs or liver. Injection of mRNA into the lymph nodes directly targets the antigen-presenting dendritic cells of the immune system, which are able to migrate from the injection site to the proximal lymph node chain.

A detailed review of the different modes of mRNA delivery and the therapeutic effects induced was finally released in 2014, in the highly cited *Nature* article "mRNA-based therapeutics—developing a new class of drugs", by Ugur Sahin, Katalin Kariko and Özlem Türeci [11] (see, in particular, the section "Progress in improving mRNA delivery").

For the global mRNA vaccine campaign against Covid, the intramuscular route was chosen by Moderna and BioNTech, following pre-marketing clinical trials of their vaccines.

Prior to 2019, few studies mentioned intramuscular injections of mRNA. The possibility of using this mode of injection, however, was raised, in 2015, by Drew Weissman in a study of mice that compared different routes of mRNA delivery for protein expression [12].

It appears that BioNTech had formulations to perform such injections as of 2017. Despite the reduced number of dendritic cells in the muscle, the immune responses induced by this mRNA delivery route have been excellent. Especially since the mechanical tearing of the tissue (by infiltration of fluid into the muscle) creates inflammation, which activates the immune system!

As journalist and physician Marc Gozlan points out: *"This route of administration is the one traditionally used for almost all vaccines. [...] Intramuscular and subcutaneous administration are the two main routes of administration of vaccines in humans. This is mainly because both are simple to administer and equally effective"* [13].

Other modes of delivery of mRNA vaccines are being investigated:

- intratumoral injection and intravenous injection, for anticancer immunotherapy (cf. *infra* part II, Chap. 13);
- intranasal vaccination: the vaccine is introduced into the body by spraying it into the nose, with a jet in each nostril, or by deposition. Intranasally-injected mRNA vaccine formulations are currently being tested in mice. Given the close proximity of the nasal cavity to the olfactory bulb—an integral part of the central nervous system—possible neurological side effects should be monitored.

- ## 5—Significant progress in cellular transfer of mRNA

The progress made in optimizing the delivery of intact mRNA to cells did not happen overnight; rather, it took twenty years!

Since its creation, CureVac has been working on the optimization of mRNA transport vectors. Thus, in 2000, to improve the transport of the mRNA molecule, CureVac proposed a lyophilization of solutions using mannose, a sugar often used for packaging (patent filed in 2011 [14]).

Aqueous solutions containing sodium, calcium and potassium salts were then developed in 2006: they increase RNA transfer and/or RNA protein translation, both in vitro and in vivo [15].

At the University of Zurich (Switzerland), Alexander Knuth and Steve Pascolo have also been working hard on this optimization. On May 26, 2009, they filed a patent protecting the formulation of protamine/RNA nanoparticles, leading to the production of immunostimulant drugs [16], according to a protocol distinct from those previously proposed by CureVac.

In this case, the most important progress was made on liposomes and cationic polymers. Positively charged, these molecules are very efficient for the encapsulation of mRNA, and for its delivery into the negatively charged cytoplasm of the cell.

Liposomes have variable mechanical and thermal characteristics, which will depend on their compositions. At room temperature, if formulations are not optimal, small liposomes tend to aggregate over time, like small soap bubbles that join together to form larger ones. As changes in the size and shape of a liposome affect its biological functionality, it is important that it maintains its size as originally designed. For example, lipofectamine appears to be a

Fig. 9.3 Chantal Pichon

very stable liposome, but it does not work very well in vivo. In the fight against Covid-19, CureVac has developed a stable liposome that retains its anti-Covid mRNA vaccine at + 4 °C; its formulation is currently protected by a patent. The liposomes used by BioNTech and Moderna appear to be less stable, which is why their Covid mRNA vaccines are stored at very low temperatures (Fig. 9.3).

Moreover, as early as 2007, Chantal Pichon, professor at the University of Orléans, France, and Patrick Midoux, Inserm research director at the CNRS, also in Orléans, developed and patented original lipids, which they exploited to make a hybrid formulation, called lipopolyplexe, composed of polymers and liposomes, capable of targeting the dendritic cells of the spleen, a major organ of the immune system, after intravenous injection [17]. This formulation has been used to deliver mRNAs encoding melanoma and head and neck cancer tumor antigens in order to induce an antitumor response (see also *infra* Part II, Chap. 13, "Experiments and Clinical Trials Conducted Against Cancer"). In collaboration with the Belgian company eTheRNA, they demonstrated that this type of formulation was completely harmless, with very little inflammation, like liposomes, and yet capable of inducing a therapeutic vaccination that completely inhibited tumor progression after implantation in the said tumor [18]. This response, obtained with both modified and unmodified RNAs (see below, Part II, Chap. 10, *"Modified and Unmodified mRNAs: For What Purpose?"*), demonstrates the non-negligible effect of the transporter in immune responses. This type of formulation, used in nasal instillation, is also effective in inducing prophylactic vaccination against the influenza virus [19]; it should also be noted that this type of formulation is currently under development for the production of anti-Covid-19 mRNA vaccines by StemiRNA Therapeutics.

Finally, since the end of the 2000s, BioNTech has been working on optimizing mRNA transfer vehicles, and, in March 2015, it proposed lipid particles capable of transporting the mRNA of interest.

More recently, in April 2019, a radioactive imaging study led by Professor Philip Santangelo demonstrated the efficacy of a specific lipid nanoparticle called CHOLK, derived from a natural sugar, and capable of carrying mRNA molecules into the cytosol of targeted cells and to lymph nodes [20].

However, the breakthrough came from Canadian biotech Acuitas Therapeutics. The company has synthesized more than 500 positively charged lipid nanoparticle vesicles that can package, protect and deliver mRNA to the cell in vivo. Clinical trials on monkeys have been successful.

The formulations used to deliver the first commercially available in vitro transcribed mRNA vaccine—BioNtech/Pfizer's Covid-19 vaccine—are based on technologies from Acuitas Therapeutics. Currently, CureVac is also using this biotech's formulations for the development of its Covid-19 mRNA vaccine.

PROF. RAMMENSEE.

References

1 Silke Holtkamp, Sebastian Kreiter, Abderraouf Selmi, Petra Simon, Michael Koslowski, Christoph Huber, Özlem Türeci, Ugur Sahin, "*Modification of antigen-encoding RNA increases stability, translational efficacy, and T-cell stimulatory capacity of dendritic cells,*" *Blood*, August 29, 2006. DOI: https://doi.org/10.1182/blood-2006-04-015024.

2 Florian Von der Mülbe, Ingmar Hoerr, Steve Pascolo, *Pharmaceutical composition containing a stabilized mRNA optimized for translation in its coding regions*, German patent (CureVac), No. WO2002098443, international filing date: 6 June 2002, publication date: 12 December 2002. Available at: https://patentscope.wipo.int/search/fr/detail.jsf?docId=WO2002098443&tab=PCTBIBLIO&_cid=P10-KML3W4-94449-1.

3 Steve Pascolo, "*Messenger mRNA-based vaccines*", *Expert Opinion Biological Therapies,* August 4, 2004, pages 1285–1294. DOI: https://doi.org/10.1517/14712598.4.8.1285.

4 Thomas Ketterer, Florian Von der Mülbe, Ladislaus Reidel, Thorsten Mutzke, *Method for preparative scale RNA purification by HPLC*, German patent (CureVac), No. WO2008077592, international filing date: 20 December 2007, publication date: 3 July 2008. Available at: https://patentscope.wipo.int/search/fr/detail.jsf?docId=WO2008077592&_cid=P11-KOOKNO-96649-1.

5 Katalin Kariko, Hiromi Muramatsu, Janos Ludwig, Drew Weissman, "*Generating the optimal mRNA for therapy: HPLC purification eliminates immune activation and improves translation of nucleoside-modified, protein encoding mRNA*", Katalin Kariko, Hiromi Muramatsu, Janos Ludwig, Drew Weissman, *Nucleic Acids Research*, November 2011. DOI: https://doi.org/10.1093/nar/gkr695.

6 Ugur Sahin, Andreas Kuhn, Edward Darzynkiewicz, Jacek Jemielity, Joanna Kowalska, *RNA-containing vaccine composition with modified 5'-cap*, German patent (BioNTech *AG, TRON,* Johannes Gutenberg University Mainz), No. CA2768600, date of application: 3 August 2010, date of publication: 10 February 2011. Available at: https://patentscope.wipo.int/search/fr/detail.jsf?docId=CA94541755&_cid=P10-KML4BB-97824-1.

7 Alexandra Orlandini Von Niessen, Stephanie Fesser, Britta Vallazza, Tim Beissert, Andreas Kuhn, Ugur Sahin, *3' utr sequences enabling RNA stabilization*, German patent (*TRON*, BioNTech), No. WO2017059902A1, international filing date: October 7, 2015, publication date: April 13, 2017. Available at: https://patentscope.wipo.int/search/fr/detail.jsf?docId=WO2017059902&_cid=P10-KML4EM-98805-3.

8 Sebastian Kreiter, Abderraouf Selmi, Mustafa Diken, Michael Koslowski, Cedrik M Britten, Christoph Huber, Özlem Türeci, Ugur Sahin, "*Intranodal vaccination with naked antigen-encoding RNA elicits potent prophylactic and therapeutic*

antitumor immunity,", *Cancer Research*, November 15, 2010; No. 70 (22); pp. 9031–40. DOI: https://doi.org/10.1158/0008-5472.CAN-10-0699.

 9 Steve Pascolo, "Emerging from the Pandemic," article published on February 17, 2021 on the Swiss Medical Forum and available at: https://medicalforum.ch/fr/detail/doi/fms.2021.08742.

10 Ugur Sahin, Heinrich Haas, Sebastian Kreiter, Mustafa Diken, Daniel Fritz, Martin Meng, Mareen Lena Kranz, Kerstin Reuter, *RNA formulation for immunotherapy*, German patent (BioNTech *AG*, *TRON*, Johannes Gutenberg University Mainz), No. EP2830593, date of application: 25 March 2013, publication date: 4 February 2015. Available at: https://patentscope.wipo.int/search/fr/detail.jsf?docId=EP130719210&_cid=P10-KML49B-97328-9.

11 Ugur Sahin, Katalin Kariko, and Özlem Türeci, *"MRNA-based therapeutics - developing a new class of drugs,"* *Nature Reviews*, October 2014, Volume XIII, pp. 759–780. DOI: https://doi.org/10.1038/nrd4278.

12 Norbert Pardi, Steven Tuyishime, Hiromi Muramatsu, Katalin Kariko, Barbara L. Mui, Ying K. Tam, Thomas D. Madden, Michael J. Hope, Drew Weissman, *"Expression kinetics of nucleoside-modified mRNA delivered in lipid nanoparticles to mice by various routes,"* Norbert, *Journal of Controlled* Release, vol. 217, 10 November 2015, Pages 345–351. DOI: https://doi.org/10.1016/j.jconrel.2015.08.007.

13 Marc Gozlan, "*L'aventure scientifique des vaccins à ARN messager*", *Le* Monde, 14 December 2020. Article available at: https://www.lemonde.fr/blog/realitesb iomedicales/2020/12/14/laventure-scientifique-des-vaccins-a-arn-messager/.

14 Thorsten Mutzke, *Mannose-containing solution for freeze-drying, transfection and/ or injection of nucleic acids*, German patent (CureVac), No. WO2011069529, international filing date: 9 December 2009, validation date: 16 June 2011. Available at: https://patentscope.wipo.int/search/fr/detail.jsf?docId=WO2 011069529&tab=PCTBIBLIO&_cid=P21-KMVUZ7-28686-4.

15 Ingmar Hoerr and Steve Pascolo, *RNA Injection Solution*, German patent (CureVac), No. WO2006122828, international filing date: May 19, 2006, publication date: November 23, 2006. Available at: https://patentscope.wipo.int/search/fr/detail.jsf?docId=WO2006122828&_cid=P10-KML3XW-948 95-1.

16 Alexander Knuth and Steve Pascolo, *Protamine/RNA Nanoparticles for Immunos-timulation*, Swiss Patent (University of Zurich, Alexander Knuth, Steve Pascolo), No. WO2009144230, International Filing Date: 26 May 2009, Publication Date: 3 December 2009. Available at: https://patentscope.wipo.int/search/fr/detail.jsf?docId=WO2009144230&_cid=P10-KML403-95418-1.

17 M. Mockey, E. Bourseau, V. Chandrashekhar, A. Chaudhuri, S. Lafosse, E. Le Cam, V F J Quesniaux, B. Ryffel, C. Pichon, P. Midoux, *"mRNA-based cancer vaccine: prevention of B16 melanoma progression and metastasis by systemic injection of MART1 mRNA histidylated lipopolyplexes,"* *Cancer gene therapy*, September 2007, vol. 14(9), p. 802–14. DOI: https://doi.org/10.1038/sj.cgt.7701072.

18 Kevin Van der Jeught, Stefaan De Koker, Lukasz Bialkowski, Carlo Heirman, Patrick Tjok Joe, Federico Perche, Sarah Maenhout, Sanne Bevers, Katrijn Broos, Kim Deswarte, Virginie Malard, Hamida Hammad, Patrick Baril, Thierry Benvegnu, Paul-Alain Jaffrès, Sander A. A. Kooijmans, Raymond Schiffelers, Stefan Lienenklaus, Patrick Midoux, Chantal Pichon, Karine Breckpot, Kris Thielemans, "*Dendritic Cell Targeting mRNA Lipopolyplexes Combine Strong Antitumor T-Cell Immunity with Improved Inflammatory Safety*", *ACS Nano*, 2018. DOI: https://doi.org/10.1021/acsnano.8b00966.

19 Federico Perche, Rudy Clemençon, Kai Schulze, Thomas Ebensen, Carlos A Guzman, Chantal Pichon, "Neutral Lipopolyplexes for *In Vivo* Delivery of Conventional and Replicative RNA Vaccine," *Molecular Therapy. Nucleic Acids*, September 6, 2019, vol. 17, pp. 767–775, DOI: https://doi.org/10.1016/j.omtn.2019.07.014.

20 Kevin E. Lindsay, Sushma M. Bhosle, Chiara Zurla, Jared Beyersdorf, Kenneth A. Rogers, Daryll Vanover, Peng Xiao, Mariluz Araínga, Lisa M. Shirreff, Bruno Pitard, Patrick Baumhof, Francois Villinger and Philip J. Santangelo, "*Visualization of early events in mRNA vaccine delivery in non-human primates via PET-CT and near-infrared imaging*,", *Nature Biomedical Engineering*, May 2019, pp. 371–380. DOI: https://doi.org/10.1038/s41551-019-0378-3.

10

Modified and Unmodified mRNA: For What Purpose?

Historically, CureVac introduced the technology of mRNA injected into the human body for the sole purpose of vaccination, either prophylactic for infectious diseases or curative for cancer. Experiments clearly showed that the injection of unmodified synthetic mRNA—derived from the transcription of the targeted DNA sequence and meeting the matching code (ATGC/AUGC)—induced a significant and perfectly controllable immune response, first in mice and then in humans. The injection was done intradermally. The vaccination objective was achieved!

As we have seen (cf. Part I, Chap. 5, "The Immune Response"), the immune response is then induced by the presence of unmodified mRNA of exogenous origin in the endosomes, interpreted by the organism as a danger signal by the intracellular PRR receptors. Their activation then leads to:

(i) the production of type I interferons, which have strong antiviral activity, and which influence the activation of the adaptive immune system in multiple ways (which are not yet well characterized)[1];
(ii) the blocking of protein synthesis.

---

[1] In addition to type 1 interferon, this is also the case for type 3 interferon. Interferons are cytokines whose impact on the immune system appears to be complex, some of them having, in particular, an inflammatory action. The types of interferons are distinguished according to the cellular receptors to which they bind. See also *below*, the box in Part II, Chap. 12 devoted to type 1 interferon, "The Role of Type 1 Interferon in Triggering the Adaptive Immune Response".

© The Author(s), under exclusive license to Springer Nature Switzerland AG 2023
J. Lemonnier and N. Lemonnier, *The Marathon of the Messenger*,
https://doi.org/10.1007/978-3-031-39300-6_10

For researchers and biotechs whose goal was to exploit mRNAs for protein replacement, the immune response and the blockage of protein synthesis were an obstacle. This played a decisive role in the development of modified mRNA by two American researchers, Katalin Kariko and Drew Weissman, in 2005. This is a development that has been much discussed in recent months as of this writing and one that needs to be understood.

When Drew Weissman, an immunologist, met Katalin Kariko, a molecular biologist, in 1997, he was working on HIV and she was working on mRNA-based gene therapies. To understand the full significance of this meeting, one must look back to the 1990s: at that time, the vast majority of immunologists were working on the HIV virus.

In the late 1990s, they began their work on mRNA based on Eli Gilboa's "classical" approach, which was based on the transfection of unmodified mRNA into dendritic cells. This was nothing new at the time; inflammation was even sought: in the abstract of the article, an immune response induced by this procedure is reported [1]. In particular, the following passage is explicit concerning the approach of the two researchers at the end of the 1990s: "*mRNA transfection also delivered a maturation signal to DC. Our results demonstrated that mRNA-mediated delivery of encoded Ag to DC induced potent primary T cell responses* in vitro. *mRNA transfection of DC, which mediated efficient delivery of antigenic peptides to MHC class I and II molecules, as well as delivering a maturation signal to DC, has the potential to be a potent and effective anti-HIV T cell-activating vaccine*" (Figs. 10.1 and 10.2).

Then, they observed that, when animal RNA was used instead of synthetic RNA or bacteria, there was no activation of immune cells in vitro. This is because mRNA is a molecule that appeared very early in the history of life, and the evolution of species has favored the appearance in higher organisms of more modified mRNAs.

Since then, Kariko and Weissman have tested the mRNA modifications. They found that it interfered with recognition by TLR 7 and 8, hence the idea of using the modifications to make non-immunostimulatory mRNAs for non-vaccine therapeutic purposes.

This is how they came up with the modified mRNA.

- *Drew Weissman and Katalin Kariko's invention: pseudouridine-modified mRNA*

The mRNA modification proposed by Drew Weissman and Katalin Kariko consists in modifying the composition of the synthetic mRNA by replacing,

Fig. 10.1 Drew Weissman

for example, one of the nitrogenous bases: uracil, U, with pseudo-uridine (or Pseudo-U).

Pseudo-uridine is a natural compound found, after transcription, in the RNA of all living beings. It results from a transformation (isomerization) of uridine (a chemical body that itself results from the attachment of uracil to a ribose sugar).

Weissman and Kariko's invention is protected by patent US8278036B2—"*RNA containing modified nucleosides and methods of use thereof*"—held by the University of Pennsylvania, and filed on August 18, 2005, in the United States. This patent became internationally valid as of March 1, 2007 (patent no. WO/2007/024708 filed with WIPO).

To circumvent this patent, not from a vaccine perspective, but to develop therapies based on non-immunostimulatory mRNAs, there are two processes:

Fig. 10.2 Katalin Kariko

- administer the unmodified synthetic RNA directly into the cytoplasm of the cell; TLR 7 receptors then do not recognize uracil dimers (i.e., uracil-uracil sequences), which are the main culprits of the inflammatory reaction;
- use molecules other than pseudo-uridine. While 2-thiouridine has been used, the main current options are Methoxy C and Methoxy U, which are used for the synthesis of modified mRNA by several companies, including the German company Ethris: these are substitution molecules for cytosine, C, and uracil, U. Ethris also exploits two other mRNA modifications: 5-iodo-uridine and 5-iodo-cytidine.

Modification with MethoxyC or MethoxyU avoids recognition of the mRNA by the TLR receptors mentioned above, and therefore does not lead to an inflammatory reaction. A number of companies use this type of "Metoxy C" or "Metoxy U" mRNA, such as, for example, the company Trilink, which has made a significant contribution to the optimization of synthetic mRNA (cf. *supra* Part II, Chap. 9, "Solutions for Optimizing mRNA"). This type of mRNA is patent-free. However, it appears to be less efficient than pseudo-uridine-modified mRNA: its synthesis is a little more difficult to achieve, and the translation of proteins is less satisfactory, which may be contrary to the objective of "protein replacement".

It is therefore easy to understand the importance of the discussion on pseudo-uridine-modified mRNA as developed by Drew Weissman and Katalin Kariko.

- *Pseudouridine-modified mRNA: what are its therapeutic applications?*

The description of patent US8278036B2—"*Modified nucleoside-containing RNA and methods of use thereof*" provides essential information about the intended use of pseudo-uridine-modified mRNA: "*This invention provides RNA, oligoribonucleotide, and polyribonucleotide molecules comprising pseudouridine or a modified nucleoside, gene therapy vectors comprising same, methods of synthesizing same, and methods for gene replacement, gene therapy, gene transcription silencing, and the delivery of therapeutic proteins to tissue in vivo, comprising the molecules. The present invention also provides methods of reducing the immunogenicity of RNA, oligoribonucleotide, and polyribonucleotide molecules*" [2].

This modification of mRNA is therefore aimed at developing a "de-immunized" mRNA. Uracil is indeed the base incorporated into the mRNA that would trigger the strongest immune reaction. The mRNA modified by pseudo-uridine would cause a much lesser inflammatory reaction, or even suppress it completely. It is therefore a priori used for non-vaccine therapeutic applications: the primary goal, then, is that the injected mRNA does not trigger an immune response. This point was confirmed in 2014 and in 2017:

- In October 2014, in the article "*m-RNA-based therapeutics—developing a new class of drugs*", published in *Nature* [3], representing a state-of-the-art review of mRNA research to date, Ugur Sahin, Katalin Kariko and Özlem Türeci describe the use of pseudo-uridine-modified mRNA very clearly: "*The recent progress in identifying RNA sensors in cells and the structural elements within mRNA that are involved in immune recognition provides opportunities to augment immune activation by IVT mRNA, or alternatively to create "de-immunized" mRNA as needed*"; and further: "*De-immunized IVT mRNA can be created by incorporating naturally occurring modified nucleosides such as pseudouridine*";
- in 2017, BioNTech and Moderna acquired sublicenses to exploit the University of Pennsylvania patent, for €75 million each, from Cellscript, the U.S. company that had the exclusive license to the patent. BioNTech's press release on the patent acquisition, dated September 6, 2017, makes it clear that the primary rationale for the purchase was not the

production of mRNAs for vaccines, but the development of therapeutic proteins: "*The patents for CELLSCRIPT's nucleoside-modified RNA technologies are the result of research conducted by Prof. Katalin Karikó, Ph.D. and Drew Weissman, M.D., Ph.D. at the University of Pennsylvania Medical School. Prof. Karikó, one of the world's leading experts in mRNA biochemistry, with more than 30 years of experience, joined BioNTech in 2013. She has published more than 70 peer-reviewed papers and conducted groundbreaking research into the discovery that incorporation of modified nucleosides suppresses immunogenicity of RNA, consequently demonstrating the feasibility of using nucleoside-modified mRNA for protein replacement* in vivo*"* [4].

The objective is therefore the translation of so-called "replacement" proteins.

Yet, a surprising change in the use of this modified mRNA occurred in this same year 2017!

In 2016, just after the emergence of a serious Zika epidemic in Latin America, and in Brazil in particular, therapeutic research focused on the development of a vaccine, in the absence of any preventive treatment or specific antiviral against the Zika virus. It was in this context that an article by Drew Weissman on the use of modified mRNA for vaccination against Zika was published in Nature at the beginning of 2017, even though this mRNA was previously considered not to provoke an immune response [5]. Protective immunity against the Zika virus is thus obtained using this modified mRNA. In particular, Drew Weissman writes: "*Here we designed a potent anti-zika virus (ZIKV) vaccine in which the premembrane and envelope glycoproteins of ZIKV are encoded by mRNA containing the modified nucleoside 1-methylpseudouridine, which prevent innate immune sensing and increases mRNA translation* in vivo*"*. This is the first reference to the still-present immunogenicity of the modified mRNA: until then, Drew Weissman had indeed claimed that this modified mRNA alone was not immunogenic!

This is a perplexing one hundred and eighty degree turn. It is essential to be able to explain it.

Beforehand, and in order to fully understand the question posed, the following table (Table 10.1) helps to identify the main objectives and areas of research, as well as the difficulties posed by the two types of mRNA. The biotechs currently working on each therapeutic application are also mentioned. The last column of the table refers to the different chapters of the

Table 10.1 The two types of mRNA—modified and unmodified

Type of mRNA	Nature of the modification	Main therapeutic applications	Main actors involved	Objective	Presupposed difficulties	Experimental stage	Clinical results	Parts of the book where the subject is mentioned
Modified synthetic mRNA	Replacement of immunostimulatory nucleotides by non-immunostimulatory nucleotides (such as Pseudo-U and MetoxyU) ?	Replacement proteins (cancer, genetic diseases, degeneratives diseases, hormonal diseases, etc.)	BioNTech, Moderna, Ethris	Development of functional proteins without induction of immune response	Inflammatory reaction	Preclinical studies Phase II clinical trials (Moderna)	Temporary effectiveness due to a possible inflammatory reaction	Part II, Chap. 14
		After 2017: infectious diseases (Zika, Covid 19, etc.)	BioNTech, Moderna	Immune response (vaccination)	Lack of inflammatory response	Common use: BioNTech and Moderna Covid vaccines	Mostly mild side effects	Part II, Chap. 12 Part III, Chap. 19

(continued)

Table 10.1 (continued)

Type of mRNA	Nature of the modification	Main therapeutic applications	Main actors involved	Objective	Presupposed difficulties	Experimental stage	Clinical results	Parts of the book where the subject is mentioned
Unmodified synthetic mRNA	–	Cancers	CureVac, BioNTech, Etherna	Immune response (vaccination)	–	Phase II clinical trials (BioNTech)	Variable and temporary efficacy—to be combined with oter cancer therapies	Part II, Chap. 6 Part II, Chap. 8 Part II, Chap. 13
		Infectious diseases (influenza, Covid-19, AIDS, etc.)	CureVac, BioNTech, Etherna	Immune response (vaccination)	–	=> Phase II clinical trials: influenza and cytomégalovirus	Mostly mild side effects	Part II, Chap. 6 Part II, Chap. 8 Part II, Chap. 12 Part III, Chap. 19

book that evoke each of the research tracks. Finally, the red arrow indicates the question posed by the change in use of modified mRNA in 2017.

- *Modified mRNA, unmodified mRNA and vaccination*

This "repositioning" by the inventors of the modified mRNA raises the question of the impact on the patient's immune system of the use of pseudouridine mRNA.
The pseudo-uridine modification makes the mRNA invisible to the intracellular receptors, called PRRs (*Pattern Recognition Receptors*), mentioned above.

As a result, the modified mRNA is no longer interpreted by the patient's immune system as a danger signal. Under these conditions, protein translation is not prevented. The desired result is then to obtain more antigens, which, a priori, would trigger a stronger immune response. However, as will be seen later (see *below*, Part II, Chap. 12, "Experiments and Clinical Trials Against Infectious Diseases"), the clinical experiments carried out lead us to question in depth the causal link mentioned here.

In addition, as discussed above, the cell transfer system used plays a role in the immune response obtained. Therefore, regardless of the type of mRNA used (modified or unmodified), it is certain that special attention will have to be paid to the formulation of lipid nanoparticles.

Under these conditions, can we continue to use unmodified mRNA for mRNA vaccination? In this field, as in others, **experiments and clinical trials** will provide valuable information.

References

1 D. Weissman, H. Ni, D. Scales, A. Dude, J. Capodici, K. McGibney, A. Abdool, S. N. Isaacs, G. Cannon, K. Karikó, "*HIV gag mRNA transfection of dendritic cells (DC) delivers encoded antigen to MHC class I and II molecules, causes DC maturation, and induces a potent human in vitro primary immune response,*" The Journal of Immunology, October 15, 2000, DOI: https://doi.org/10.4049/jimmunol.165.8.4710.
2 Katalin Kariko, Drew Weissman, *RNA containing modified nucleosides and methods of use thereof,* U.S. Patent (University of Pennsylvania), No. US8278036B2, publication date: August 23, 2005. Available at: https://patents.google.com/patent/US8278036B2/en.

3 Ugur Sahin, Katalin Kariko, and Özlem Türeci, "*MRNA-based therapeutics - developing a new class of drugs,*" *Nature Reviews*, October 2014, Volume XIII, pp. 759–780. DOI: https://doi.org/10.1038/nrd4278.
4 Press release available at: https://www.globenewswire.com/news-release/2017/09/06/1108043/0/en/BioNTech-AG-Enters-into-Licensing-Agreement-with-CEL LSCRIPT-LLC-as-it-Advances-Development-of-Messenger-RNA-Encoding-Bis pecific-Antibodies-and-other-Therapeutic-Proteins.html.
5 Norbert Pardi, Michael J. Hogan, Rebecca S. Pelc, Hiromi Muramatsu, Hanne Andersen, Christina R. DeMaso, Kimberly A. Dowd, Laura L. Sutherland, Richard M. Scearce, Robert Parks, Wendeline Wagner, Alex Granados, Jack Greenhouse, Michelle Walker, Elinor Willis, Jae-Sung Yu, Charles E. McGee, Gregory D. Sempowski, Barbara L. Mui, Ying K. Tam, Yan-Jang Huang, Dana Vanlandingham, Veronica M. Holmes, Harikrishnan Balachandran, Sujata Sahu, Michelle Lifton, Stephen Higgs, Scott E. Hensley, Thomas D. Madden, Michael J. Hope, Katalin Karikó, Sampa Santra, Barney S. Graham, Mark G. Lewis, Theodore C. Pierson, Barton F. Haynes, Drew Weissman, "*Zika virus protection by a single low-dose nucleoside-modified mRNA vaccination,*" *Nature*, 9 March 2017, vol. 543(7644), pp. 248–251. DOI: https://doi.org/10.1038/nature21428.

11

Experiments and Clinical Trials Conducted: The Power of Therapeutic mRNA

Since medicine is, above all, in addition to an art, an experimental science, therapeutic research necessarily implies the carrying out of clinical trials. In his *Histoire de la pensée médicale*, Maurice Tubiana clearly and precisely indicates how these trials are generally conducted[1]:

"*In the evaluation of a new drug, the first step is experimental, first in vitro, then in animals, which is essential to detect toxicities (especially carcinogenic and teratogenic effects). If the results in two animal species are satisfactory, the clinical stage begins. This is carried out in three phases. The first phase verifies the absence of toxicity in humans and begins with doses lower than those expected to have a therapeutic effect. This phase is essential because animal testing is not sufficient to exclude the risk of toxic effects in humans. In phase II, the product is tested on a small number of patients (a few dozen) to estimate its efficacy at various doses and to determine the optimal doses. One might think that patients would be reluctant to try drugs whose action is unknown. On the contrary, at an advanced stage of the disease, these trials are a source of comfort for patients who see them as a new chance. It is only after phase II that the therapeutic trial is launched on hundreds of patients to compare the new drug with the conventional treatment [phase III]).*"

---

[1] Cf. Maurice Tubiana, *Histoire de la pensée médicale. Les chemins d'Esculape, op. cit.* p. 311.

J. Lemonnier and N. Lemonnier, *The Marathon of the Messenger*, https://doi.org/10.1007/978-3-031-39300-6_11

When clinical trials conclude that a new molecule is effective, a number of elements that are essential for its marketing must be taken into account:

(i) its possible toxic effects in the more or less long term, which implies assessing the risks on a case-by-case basis, since there is no universal grid for assessing the risks incurred in all situations, whatever the protocol followed;

(ii) the financial and social cost of producing this new product, which is also part of the analysis of the improvement of the service rendered to the patient in comparison with one or more treatments already marketed;

(iii) the often long and complex procedures required for marketing authorization, with patient safety as the primary concern.

Preclinical research, conducted on animals, results in many failures. Only five percent of the molecules tested in preclinical trials "go on" to clinical trials, after which perhaps only 1% of the molecules will be retained.

Clinical trials are not predefined by beliefs, ideologies, knowledge, opinions or even personal intuitions. They are carried out in practice without any preconceived ideas, in all directions. As a result, if the experimental approach leads to a certain number of errors, it also leads to initially unsuspected discoveries. This is what we call serendipity, e. g., the experience of making an unexpected scientific discovery, following various experiments, and then seizing all the practical significance of it.

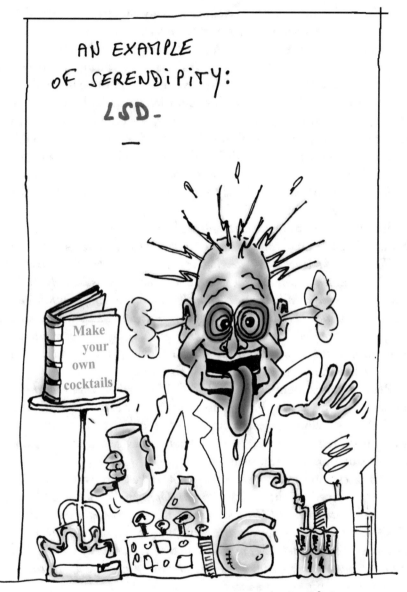

This is how some research avenues initially perceived as secondary, dead-end and even possibly bad, such as mRNA vaccines in the early 2000s, finally turned out to be, twenty years later, real therapeutic solutions.

To be fully reliable in humans, clinical trials must include people of all ages, including those over 75. For vaccines, volunteers must be in good health and not be taking medication (which could directly or indirectly interfere with the experimental treatment). This excludes, for example, pregnant women, nursing mothers, people with autoimmune diseases, etc. It should also be noted that, in practice, more difficulties can arise if you have a large number of elderly volunteers in the clinical test phase.

All these steps were obviously essential in the development of the anti-Covid mRNA vaccines.

Specifically regarding experiments and clinical trials that use mRNA, Ugur Sahin, Özlem Türeci, and Katalin Kariko wrote, in 2014, in their article "*MRNA-based therapeutics—developing a new class of drugs*" in *Nature Reviews* [1], "*As outlined in this Review, cancer immunotherapy is the only field in which clinical testing and industrialization of the manufacturing of mRNA drugs is at an advanced stage. For vaccination against infectious diseases, IVT mRNA is in early clinical testing, whereas in all other medical applications, such as protein replacement, it is at preclinical stage*". The Covid pandemic of 2020 helps to measure how far we have come in just six years.

Recall that, as mentioned in §II.3 ("*The birth of CureVac: the era of the pioneers*"), Hans-Georg Rammensee's vision, as early as 1996, was to use mRNA to develop cancer vaccines. In the context of the 2000s, mRNA was a basic research topic; 20 years later, it has become a field for clinical experimentation and treatment.

It is interesting to note the view of Steve Pascolo, who states, in a 2004 article published in the journal *Expert Opinion on Biological Therapy* [2]: "*Increased interest in mRNA vaccination as a safe and efficient replacement for DNA-based vaccines is expected to result in a larger number of human trials, regardless of which method of mRNA delivery (...) would be used (...) The extension to human therapy using these very promising and safe approaches requires a large amount of mRNA produced according to GMP guidelines. [...]*").

And, three years later, his team reported, in the journal *Gene Therapy*, the experimental evidence that definitively opens the way to mRNA vaccines [3]. These researchers synthesized in vitro a synthetic mRNA coding for luciferase, a reporter protein.[2] They then injected this mRNA, placed in a suitable saline

---

[2] Experiments conducted by Robert W. Malone at the Salk Institute of Biological Studies had, in 1989, demonstrated the expression of luciferase after in vitro transfection of mRNA in mouse cells.

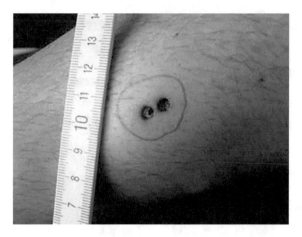

Fig. 11.1 mRNA injection point (knee)

medium that responds to GMP conditions, into the dermis of a man (Steve Pascolo himself).

A few hours later (the experiment having been repeated three times), skin biopsies were taken that, incubated in the presence of luciferin, emitted bioluminescence. In other words, the synthetic mRNA was taken over by the translational machinery of the skin cells, and a functional luciferase catalyzed the luciferin: the synthetic mRNA is indeed functional in vivo in humans! Thus, there is no need to transfect the mRNA onto dendritic cells before 1reinjecting them into the patient! Moreover, the skin at the site of injection of the synthetic mRNA is neither inflamed nor necrotic, and Pascolo did not experience any systemic side effects after these multiple injections. This demonstrates that the unmodified synthetic mRNA is not toxic (neither locally nor systemically) (Fig. 11.1).

This makes things easier: clinical trials based on direct injection of mRNA can begin.

By the end of the 2000s, preclinical and clinical trials had confirmed that synthetic mRNA is the option of the future.[3]

However, the synthetic mRNA is transfected here; it would be used "naked" one year later, in 1990, in the study of Wolff and *al* mentioned in references below [4].

[3] Itziar Gómez-Aguado *et alii's article in Nanomaterials,* published on February 20, 2020, brings together the experimental approaches and achievements obtained with synthetic mRNA [5].

Inadvertent discovery of nuclear fission
– which also lead to atomic bomb –

References

1 Ugur Sahin, Katalin Kariko, and Özlem Türeci, *"MRNA-based therapeutics - developing a new class of drugs,"*, Nature Reviews, October 2014, volume XIII, pp. 759–780. DOI: https://doi.org/10.1038/nrd4278.

2 Steve Pascolo, *"Messenger mRNA-based vaccines,"* Expert Opinion Biological Therapies, August 4, 2004, pages 1285–1294. DOI: https://doi.org/10.1517/147 12598.4.8.1285.

3 J. Probst, B. Scheel, B.J. Pichler, I. Hoerr, H-G Rammensee, S. Pascolo, *"Spontaneous cellular uptake of exogenous messenger RNA in vivo is nucleic acid-specific, saturable and ion-dependent,"* Gene Therapy (2007) 14: 1175–1180. DOI: https://doi.org/10.1038/sj.gt.3302964.

4 Jon A. Wolff, Robert W. Malone, Phillip Williams, Wang Chong, Gyula Acsadi, Agnes Jani, Philip. L. Felgner, *"Direct Gene Transfer into Mouse Muscle in Vivo,"* Science, March 23, 1990, vol. 247. DOI: https://doi.org/10.1126/science.169 0918.

5 Itziar Gómez-Aguado, Julen Rodríguez-Castejón, Mónica Vicente-Pascual, Alicia Rodríguez-Gascón, María Ángeles Solinís, Ana del Pozo-Rodríguez, *"Nanomedicines to Deliver mRNA: State of the Art and Future Perspectives,"* Nanomaterials, 20 February 2020. DOI: https://doi.org/10.3390/nano10020364.

12

Experiments and Clinical Trials Against Infectious Diseases

Let's remember two essential steps:

- 1993: the seminal work by Pierre Meulien's team at the Hôpital Cochin, Paris: this was the first demonstration of an immune response in mice after an injection of synthetic mRNA. In this world's first, the mRNA encoded an influenza virus protein *(see Part II*, Chap. 6, "Promising Studies from the 1990s").
- 2000: ignoring the general skepticism of the scientific world at the time, CureVac took up this concept *(see Part II*, Chap. 8, "The Birth of CureVac: The Era of the Pioneers") by stating its *raison d'être*: a biotech that uses synthetic mRNA to treat certain cancers and infectious diseases.

The optimization of mRNA vaccines against certain infectious diseases took about 20 years, culminating in its success today in quelling the Covid-19 pandemic.

Currently, in vitro transcribed mRNA vaccines have advantages, which were of particular interest when the first clinical trials were conducted.

- Rapid production under pharmaceutical conditions

Once its production process has been established, any synthetic mRNA can be made and, therefore, can code for any protein within a few weeks. For

© The Author(s), under exclusive license to Springer Nature Switzerland AG 2023
J. Lemonnier and N. Lemonnier, *The Marathon of the Messenger*, https://doi.org/10.1007/978-3-031-39300-6_12

viruses that mutate relatively easily, this ensures a much faster development of vaccines adapted to the various variants.

- A safe and defined therapeutic window

Since the mRNA molecule is rapidly degraded in the body's cells, the unmodified or modified mRNA, degraded by RNase enzymes, disappears from the cell a few days after injection. The mRNA, even if not rapidly degraded, could not alter the DNA genome of every human being anyway. Therefore, the side effects induced by the mRNA vaccine are transient and allow individuals to have further injections to ensure lasting efficacy of the treatment, without fear of accumulation of the therapeutic product (which is not the case when DNA is injected, as in Astrazeneca's Covid-19 vaccine).

- A targeted response against a single antigen

mRNA vaccines produce one and only one immune response, because they encode only one viral antigen (e.g., the "spike" envelope protein of Sars-Cov-2). Since they contain no potentially immunogenic contaminating proteins, they are very pure. Finally, they are formulated without other antigens. Depending on the viruses targeted, it appears nevertheless possible to target several distinct proteins, if this can improve the overall immune response or help design an mRNA vaccine against different viruses (influenza, Covid-19…) or different variants of the same virus.

In 2008, Steve Pascolo was already aware of the interest in such an mRNA vaccination against infectious diseases, when he wrote [1]: "*The naturally transient and cytosolically active mRNA molecules are seen as a possibly safer and more potent alternative to DNA for gene vaccination. Optimized mRNA (improved for codon usage, stability, antigen-processing characteristics of the encoded protein,* etc.) *were demonstrated to be potent gene vaccination vehicles when delivered naked, in liposomes, coated on particles or transfected in dendritic cells* in vitro. *Human clinical trials indicate that the delivery of mRNA naked or transfected in dendritic cells induces the expected antigen-specific immune response*".

The mRNA used for these first clinical trials is naturally unmodified by pseudo-uridine.

As noted above (see *supra* Part II, Chap. 10, "Modified and Unmodified mRNA: For What Purpose?"), modified mRNA, especially that modified by pseudouridine, produces more protein than unmodified mRNA in immune cells, because type 1 interferon, induced by unmodified mRNA, blocks protein expression in most cells (see *supra* Part I, Chap. 5, "The Immune

Response").

In this case, type 1 interferon influences the activation of the adaptive immune response and, in doing so, plays a favorable role for unmodified mRNA vaccination. Indeed, as detailed in the box below, some specialized cells of the immune system are resistant to type 1 interferon and produce the proteins they present to the immune system. This is essential for the induction of adaptive antiviral immune responses, and therefore for preventing the fatal outcome of the infection: antigen presentation will thus continue to take place, even using unmodified mRNA! Unlike mRNA used to compensate for a defect or the absence of a protein of interest ("protein replacement"), vaccination does not require large amounts of protein; it is just necessary that it be effectively presented by dendritic cells to cytotoxic B and T lymphocytes.

12.1 The Role of Type 1 Interferon in Triggering the Adaptive Immune Response

Type 1 interferon induced by unmodified mRNA blocks protein expression, but also triggers activation of the adaptive immune response; thus, it plays two opposing roles in the context of unmodified mRNA vaccination. The question then is: of these two roles—positive effect on the immune response and negative effect on antigen production—which is dominant?

Two convincing studies provide some answers on this subject:

1. first, the study by Karl S. Lang's team [2], "Enforced viral replication activates adaptive immunity and is essential for the control of a cytopathic virus", published in *Nature Immunology* in 2011, mentions that "*the innate immune system limits viral replication* via *type I interferon and also induces the presentation of viral antigens to cells of the adaptive immune response*". To understand what we are talking about, let's first recall that, upon injection of an unmodified exogenous mRNA, the type I interferon response has the effect of blocking the translation of the viral proteins that are the antigens. However, in certain specialized macrophages that are present in the marginal zone of the spleen, and which play an important role in pathogen resistance, protein production continues precisely in order to stimulate the adaptive immune system. The authors of the study "*found that expression of the gene encoding the inhibitory protein Usp18 in metallophilic macrophages resulted in reduced responsiveness to type I interferons, allowing locally restricted replication of the virus.*" As a result, some specialized cells of the immune system are resistant to interferon and produce the

proteins they present to the immune system: this is critical "*for the induction of adaptive antiviral immune responses and, therefore, for preventing the fatal outcome of the infection.*" Antigen presentation will therefore continue to take place, even when using unmodified mRNA. This immunological mechanism will thus ensure efficient activation of the adaptive immune response, an activation sought for vaccination;

2. the recent study by Gerardo Guillén-Nieto's team [3], "*Revisiting Pleiotropic Effects of Type I Interferons: Rationale for Its Prophylactic and Therapeutic Use Against Sars-CoV-2*", published on March 26, 2021, in the journal *Frontiers in Immunology*, insists that type 1 interferon is considered to have a positive effect on T-cells (cytotoxic and helper cells) during the initial phase of viral antigen recognition: "*IFN signaling is context-specific (…), thus in virally infected cells type I IFN signaling enhances the susceptibility to undergo apoptosis, thereby, preventing viral replication and spread (…). Dendritic cells (DC) response to type I IFN consists of their activation and secretion of proinflammatory cytokines that lead to activation of the adaptive immune response.*" (…) "*and pDCs, which secrete extremely high levels of type I IFNs, promote B cells activation and the subsequent production of antiviral antibodies*". In these circumstances, these dendritic cells "*secrete extremely high levels of type I IFNs, and promote B cell* [i.e., B cell] *activation and the subsequent production of antiviral antibodies.*" **The same article further notes that type 1 interferon is a relevant treatment for Sars-Cov-2. It emphasizes that an appropriate interferon response in fine protects against any risk of uncontrolled inflammation**, "*It also means that the antiviral and immunomodulatory effects of IFNs synchronized with other IFNs' benefits for the direct or indirect control of inflammatory cytokines, neutrophilia, regulatory T cells, and the induction of ACE2 expression, may help to mimic a physiological antiviral response, with an intact IFN signaling system*".

It should also be noted that BioNTech uses unmodified mRNA in its cancer vaccine clinical trials. A type 1 interferon response, as well as the triggering of the adaptive immune response (T lymphocytes), is then precisely sought.

More broadly, the real question posed today by the use of modified or unmodified mRNA is that of the immune response induced. In this regard, an important study, conducted in 2018 by Drew Weissman's team, demonstrated that "*modified mRNA vaccines induce potent immune responses in helper T cells and B cells*" [4]. It is reported that "*comparative studies demonstrated that nucleoside-modified mRNA-LNP vaccines outperformed adjuvanted protein and inactivated virus vaccines and pathogen infection. The incorporation of nonin-flammatory, modified nucleosides in the mRNA is required for the production of large amounts of antigen and for robust immune responses".*

However, it has been found that the use of unmodified mRNA in various clinical experiments can also induce a strong antibody response. As early as the end of the 2000s, clinical trials against cancer also reported a strong antibody response [5]; this will be specifically addressed in the next section.

On the preclinical level, for example, we should mention the work carried out by Benjamin Petsch and his colleagues at the Friedrich-Loeffler-Institut (Tübingen): in 2012, they reported that an unmodified synthetic mRNA vaccine was capable of conferring protection against the influenza virus. Its ability to induce an immune response, in mice, ferrets and pigs, is comparable to that observed with a conventional inactivated influenza vaccine [6].

And in a phase I clinical trial published in 2017, CureVac reported that a synthetic unmodified mRNA rabies vaccine induces antibodies against rabies virus in patient volunteers [7].

Thus, since before the appearance of Covid-19, experiments and clinical trials on humans have already been conducted with unmodified mRNA to treat various infectious diseases of viral origin.

These experiments all demonstrate a very good control of the inflammatory reaction induced by intracellular mRNA addressing.

This last question was, moreover, specifically approached in 2015, during preclinical experiments by Bruno Pitard, research director at CNRS, who developed a novel unmodified mRNA delivery system with a very well-controlled immune response [8]. The study, performed on *cynomolgus* monkeys, demonstrated that the delivery of unmodified mRNA to antigen-presenting cells, both at the injection site and in the lymph nodes, induces the activation of the adaptive immune system, without triggering an inflammatory reaction.

These arguments therefore confirm the fully functional character of an unmodified mRNA vaccination, without producing an uncontrolled inflammatory reaction that would then be responsible for the death of mice, if not of human beings!

Note that Drew Weissman himself, who has insisted, on several occasions, on the serious nature of the inflammatory reactions linked to the use of unmodified mRNA, nevertheless reported, in 2017, on preclinical trials that demonstrated the possibility of using unmodified mRNA for vaccination, thanks to optimization of the adjuvant activity. Indeed, it appears possible to obtain an immune response (induced by the expression of an antigen) *"from a naked, unmodified, sequence-optimized mRNA"* [9]. Note that CureVac's work is mentioned here by an American scientist, proof, if any were needed, that it was not ignored by the researchers at the University of Pennsylvania...

In this case, some experiments on humans with mRNA vaccines (modified or unmodified) were not conclusive. Thus, in 2012–2013, trials were conducted against AIDS, without any antiviral action being observed.

- **A new type of mRNA vaccine: self-replicating mRNA vaccines**

To understand the interest in this new type of vaccine, it is first necessary to describe what self-replicating mRNA is.

As explained by the physician and journalist Marc Gozlan [10], "*this last type of RNA has the ability to self-replicate, because it encodes an enzyme called replicase. Once in the cytoplasm of the cell, the self-replicating RNA translates the replicase gene. This enzyme then copies the long RNA molecule into a complementary strand. This is then used by the replicase to make a large amount of self-replicating RNA*", which will then synthesize a large amount of viral antigens by translation much faster. These self-replicating vaccines, restricted, for the moment, to the veterinary field, cost significantly less than traditionally produced vaccines.

Even before Covid-19, some experiments had been conducted against zoonoses, infectious animal diseases that can sometimes be transmitted to humans. For example, BioNTech used self-replicating mRNA in the development of these mRNA vaccines for animals. This type of mRNA vaccine is unique in that it contains a sequence of an alphavirus or a flavivirus, which are two types of single-stranded RNA viruses.

Already, these self-replicating mRNA vaccines have been tested against zika, cytomegalovirus and AIDS.

In the fight against Covid-19, the human vaccine has so far been a conventional mRNA, although BioNTech has also developed, with Pfizer, an anti-Covid vaccine that uses self-replicating mRNA (BNT162c2 vaccine). **However, clinical trials conducted by Imperial College London and the U.S. firm Arcturus Therapeutics demonstrate the full functionality of this type of vaccine.**

On April 21, 2022, the scientific community learned of the successful development by the American company *Arcturus Therapeutics* of a third mRNA anti-Covid vaccine, using a self-replicating mRNA: the success of the phase 3 clinical trial of this vaccine conducted in Vietnam on 17,000 patients, reported by the journal *Science* [11], makes it possible to envisage its marketing in the near future.

This vaccination process is based on an enzyme found in the above-mentioned alphaviruses and flaviviruses, and is inspired by a vaccine process identified by Zhou et al. in 1994 [12], and already used by BioNTech to develop vaccines for veterinary use. The self-replicating mRNA translates the gene of the replicase enzyme. This enzyme then copies the mRNA molecule into a complementary strand. The large amount of mRNA produced in this way will allow for a much faster translation of viral antigens, in some way mimicking a viral infection. The proteins are then expressed over a longer period of time than with a non-self-amplifying mRNA vaccine, especially a modified one.

This is consistent with a more durable immune response to further exposure to the virus.

Deborah Fuller, an immunologist at the University of Washington School of Medicine, states, in the previously cited *Science* article by Jon Cohen [11], that "*a self-amplifying mRNA vaccine against COVID-19 would ideally replace the two primary doses, giving it an even clearer advantage over its conventional relatives. A booster months later might still be warranted, as is now recommended for current mRNA vaccines.*" Professor Fuller finally points out that "*self-amplifying mRNAs could also lead to more durable immune responses*".

References

1 Steve Pascolo, "*Vaccination with messenger RNA (mRNA)*," *Handbook of Experimental Pharmacology*. 2008, vol. 183, p. 221–35. DOI: https://doi.org/10.1007/978-3-540-72167-3_11.

2 Nadine Honke, Namir Shaabani, Giuseppe Cadeddu, Ursula R. Sorg, Dong-Er Zhang, Mirko Trilling, Karin Klingel, Martina Sauter, Reinhard Kandolf, Nicole Gailus, Nico van Rooijen, Christoph Burkart, Stephan E. Baldus, Melanie Grusdat, Max Löhning, Hartmut Hengel, Klaus Pfeffer, Masato Tanaka, Dieter Häussinger, Mike Recher, Philipp A Lang, Karl S Lang, "*Enforced viral replication activates adaptive immunity and is essential for the control of a cytopathic virus*", *Nature Immunology*, 20 November 2011, vol. 13(1), p. 51–7. DOI: https://doi.org/10.1038/ni.2169.

3 Diana Garcia-del-Barco, Daniela Risco-Acevedo, Jorge Berlanga-Acosta, Frank Daniel Martos-Benítez, and Gerardo Guillén-Nieto, "*Revisiting Pleiotropic Effects of Type I Interferons: Rationale for Its Prophylactic and Therapeutic Use Against Sars-CoV-2,*", *Frontiers in Immunology*, March 26, 2021. DOI: https://doi.org/10.3389/fimmu.2021.655528.

4 Norbert Pardi, Michael J Hogan, Martin S Naradikian, Kaela Parkhouse, Derek W Cain, Letitia Jones, M. Anthony Moody, Hans P Verkerke, Arpita Myles, Elinor Willis, Celia C. LaBranche, David C. Montefiori, Jenna L Lobby, Kevin O Saunders, Hua-Xin Liao, Bette T Korber, Laura L Sutherland, Richard M Scearce, Peter T Hraber, István Tombácz, Hiromi Muramatsu, Houping Ni, Daniel A Balikov, Charles Li, Barbara L Mui, Ying K Tam, Florian Krammer, Katalin Karikó, Patricia Polacino, Laurence C Eisenlohr Thomas D Madden, Michael J Hope, Mark G Lewis, Kelly K Lee, Shiu-Lok Hu, Scott E Hensley, Michael P Cancro, Barton F Haynes, Drew Weissman, "*Nucleoside-modified mRNA vaccines induce potent T follicular helper and germinal center B cell responses*," The Journal of Experimental Medicine, June 4, 2018, vol. 215 (6), p. 1571–1588. DOI: https://doi.org/10.1084/jem.20171450.

5 Benjamin Weide, Jean-Philippe Carralot, Anne Reese, Birgit Scheel, Thomas Kurt Eigentler, Ingmar Hoerr, Hans-Georg Rammensee, Claus Garbe, Steve Pascolo, "*Results of the first phase I/II clinical vaccination trial with direct injection of mRNA*,", Journal of Immunotherapy, February- March 2008, vol. 31(2), p. 180–8. DOI: https://doi.org/10.1097/CJI.0b013e31815ce501.

6 B. Petsch et al, "*Protective efficacy of in vitro synthesized, specific mRNA vaccines against influenza A virus infection*," Nature Biotech, December 2012, vol. 30 (12), p. 1210–1216. DOI: https://doi.org/10.1038/nbt.2436.

7 Martin Alberer, Ulrike Gnad-Vogt, Henoch Sangjoon Hong, Keyvan Tadjalli Mehr, Linus Backert, Greg Finak, Raphael Gottardo, Mihai Alexandru Bica, Aurelio Garofano, Sven Dominik Koch, Mariola Fotin-Mleczek, Ingmar Hoerr, Ralf Clemens, Frank von Sonnenburg, "*Safety and immunogenicity of a mRNA rabies vaccine in healthy adults: an open-label, non-randomized, prospective, first-in-human phase 1 clinical trial*,", Lancet, September 23, 2017, no. 390(10101), pp. 1511–1520. DOI: https://doi.org/10.1016/S0140-6736(17)31665-3.

8 Bruno Pitard, *Capped and uncapped RNA molecules and block copolymers for intracellular RNA delivery*, French patent (University of Nantes), No. WO2015151048, international filing date: 1er April 2015, publication date: 8 October 2015. Available at: https://patentscope.wipo.int/search/fr/detail.jsf?docId=WO2015151048&_cid=P10-KM898W-81900-1.

9 N. Pardi, M. J. Hogan, F. W. Porter, D. Weissman, "*mRNA vaccines - a new era in vaccinology*," N. Nature Review Drugs Discovery, April 2018, no. 17 (4), pp. 261–279 (2018). DOI: https://doi.org/10.1038/nrd.2017.243.

10 Marc Gozlan, "The scientific adventure of messenger RNA vaccines," article dated December 14, 2020, available on Marc Gozlan's Internet blog at: https://www.lemonde.fr/blog/realitesbiomedicales/2020/12/14/laventure-scientifique-des-vaccins-a-arn-messager/, *op. cit.*

11 Jon Cohen, *"A mRNA vaccine with a twist-it copies itself-protects against Covid 19,"* *"A mRNA vaccine with a twist-it copies itself-protects against COVID-19"* in *Science*, 21 April 2022. Article available at https://www.science.org/content/art icle/mrna-vaccine-twist-it-copies-itself-protects-against-covid-19.

12 X. Zhou, P Berglund, G Rhodes, S E Parker, M Jondal, P Liljeström, *"Self-replicating Semliki Forest virus RNA as recombinant vaccine,"* *Vaccine*, 12 December 1994, 1510–14. DOI: https://doi.org/10.1016/0264-410x(94)900 74-4.

13

Experiments and Clinical Trials Against Cancer

For cancer vaccines, the objective is not prophylactic, but therapeutic: to offer an immunization that activates immune defenses against tumors and metastases.

Through specific changes in their genome, cancer cells express specific proteins that are recognized by the body as antigens. The goal of the mRNA vaccine is for the patient to manufacture his or her own tumor antigens in specialized immune cells (antigen-presenting cells) to stimulate the immune system and help eliminate the cancer cells and reduce the size of the tumor, and even prevent recurrence. This therapy is part of a multidisciplinary approach that includes surgery, chemotherapy and radiation therapy.

CureVac, in collaboration with the Tübingen Clinic, began clinical trials on melanoma in 2003. Indeed, this type of skin cancer, which is particularly aggressive, is among those that respond best to an activation of the immune system. The melanoma vaccine was developed by extracting mRNA from the patient's own cancer cells. This synthetic mRNA is amplified in vitro (through conversion to DNA and in vitro transcription) and then injected back into the patient. This method is described in an article published by Steve Pascolo's team [1] in *Genetic Vaccines and Therapy*. In the early 2000s, German law authorized this type of autologous approach, in which the donor and the recipient are the same person; Germany, at the time, enjoyed greater freedom of medical experimentation than other countries, particularly France.

© The Author(s), under exclusive license to Springer Nature
Switzerland AG 2023
J. Lemonnier and N. Lemonnier, *The Marathon of the Messenger*,
https://doi.org/10.1007/978-3-031-39300-6_13

The results of this first clinical trial against melanoma were published in February 2008, in the *Journal of Immunotherapy* [2].

Thus, the first studies conducted on melanoma patients, confirmed between 2007 and 2010 on those with renal cell carcinoma, show, in clinical practice, the safety of unmodified mRNA vaccine formulations and the reproducibility of immune responses without uncontrolled inflammatory reactions. In particular, the results of these different trials confirm the efficacy of mRNA vaccines on metastatic melanoma with a small primary tumor.

In Tübingen, from 2003 to 2006, the quality of the clinical trials conducted by CureVac improved.

At the University Hospital of Zürich, Thomas Kündig, Chief Physician in the Department of Dermatology, and Alexander Knuth, Head of the Department of Oncology, were interested in the work of CureVac in Tübingen. At that time, Steve Pascolo wanted to improve the efficacy of the cancer vaccine by injecting mRNA into the lymph nodes to obtain a strong immune response. In Zürich, this type of injection was routinely performed by Kündig's team. Pascolo set up a collaboration between CureVac and Zurich in 2005, with two objectives: (i) a clinical trial of an anti-cancer mRNA vaccine in Switzerland (which would indeed be set up), thus independent of the European Union and (ii) the possibility to develop—preclinically and clinically—mRNA injections in lymph nodes. However, after the departure of its scientific director, CureVac decided to support the Zürich clinical trial by excluding Steve Pascolo from this project. Nor did it support his research in Zürich with the mRNA vaccine in lymph nodes. The relationship between Pascolo and CureVac was completely severed between 2006 and 2008, at the initiative of the Tübingen biotech.

The clinical trials, conducted in Zurich from 2008 to 2010 on lung cancer patients, used the CureVac techniques practiced in 2006 in Tübingen, with intradermal injections performed repeatedly, up to nine times a week. Under these conditions, the clinical trial does not appear to have been a source of comfort for the patients. The results themselves were disappointing, even according to Steve Pascolo, who is not one of the authors of the 2019 article on the study [3].

Back in Tübingen, in the wake of the 2008 financial crisis, CureVac was suffering, and research on mRNA was stalled. However, this biotech, a pioneer in the conduct of cancer clinical trials based on unmodified mRNA, was, at that time, still the only company in this niche of synthetic mRNA vaccines, and therefore had no competitors.

In this capacity, Benjamin Weide, Claus Garbe, Hans-Georg Rammensee and Steve Pascolo published an article in *Immunology Letters on* January 15, 2008, summarizing preclinical and early clinical studies on certain cancers.

This opened up promising prospects for effective immunotherapies targeted at the specific immune response of each patient.

In Mainz, Germany, BioNTech, its researchers and physicians have been working on anti-cancer mRNA vaccination since the company was founded in 2008 (and, in fact, long before that, in Ugur Sahin's academic laboratory). Several patents attest to this [5–7]. Again, as with CureVac, the goal is to target antigens that result from genetic abnormalities in certain cancers to elicit a therapeutic immune response in humans. The advances are more technical than clinical, as shown by the patents filed. In general, it appears that the cellular mutations that cause cancerous tumors to develop are effectively targeted by BioNTech's vaccines. The preclinical results are excellent and the clinical results—all published in high impact journals (*Nature*)—are extremely promising.

Starting in the second decade of the twenty-first century, in his trials, Dr. Ugur Sahin became concerned with making injections better, not simply increasing the number of injections. In particular, a first clinical trial is taking place to treat metastatic melanoma. All clinical (and preclinical) trials with BioNTech's cancer vaccines use unmodified mRNA; BioNTech teams are pleased to have induced type 1 interferon AND a strong T response in mice and patients [8, 9]. This generates a double therapeutic effect of the unmodified RNA vaccine!

For more information on this topic, please see the box entitled "Additional information on cancer trials and experiments".

Currently, several cancer vaccine trials using in vitro transcribed mRNA are underway. Ultimately, optimization of the design of in vitro transcribed mRNA (e.g., by expression of specific oncogene mutations or amplification, specific to each patient's cancer), its formulation (e.g., in intravenously administered liposomes), and its use (e.g. in combination with other cancer therapies) will certainly lead to the validated efficacy of cancer mRNA vaccines.

Finally, we should mention that clinical trials have been carried out to generate the production of cytokines (mainly interleukin 2 and interferon α) or dendritic cell activators (particularly in Etherna) from mRNA. The injection is done intravenously or intratumorally and aims to produce these proteins, which, as we have seen, play an important role in innate immunity and in the activation of helper T-cells. The objective is indeed to reactivate the immune response against cancers. However, in so far as we wish, at all costs, to avoid an inflammatory reaction in the organism, we are working within the context of a *protein replacement* approach, using "de-immunized" mRNA: see the section *below* devoted to clinical trials in this field. However, Etherna uses unmodified mRNA for its intratumoral injections.

13.1 Additional Information on Experiments and Clinical Trials Carried Out Against Cancers

In all anti-cancer clinical trials using mRNA vaccination or other types of vaccines, the key role played by the immune system in controlling tumor proliferation as completely as possible must be emphasized. Current immunotherapies are not very specific. By increasing a global immune response, they run the risk of developing autoimmunity.

Cancer vaccines aim to strengthen the immune system to kill cancer cells. The aim is to induce an immune response via the mRNA specific to the genetics of the tumor tissue. For this purpose, sequencing of the primary tumor, or even its metastases, leads to the identification of mutated or deregulated genes, in particular, those that are induced or overexpressed, a step that helps to define the mRNAs to be used in the vaccine. These vaccines induce a targeted immune response against oncogenic proteins specifically (over)expressed in cancers. This represents a fourth therapeutic option, personalized to the patient's cancer.

BioNTech's work has strengthened CureVac's new cancer treatment. In leading BioNTech, Ugur Sahin and Özlem Türeci, both clinicians and oncologists by training, have conducted pivotal clinical trials. The study published on July 13, 2017, in the journal *Nature*, "*Personalized RNA mutanome vaccines mobilize poly-specific therapeutic immunity against cancer*" [9], provided compelling clinical results for melanoma; they focus not on the primary tumor, but on metastases. A "T-cell" immune response after RNA vaccination was observed in all five patients in the clinical trial.

This approach requires the identification of mutations in the cancer prior to the manufacture of a customized vaccine mRNA. The treatment is currently in Phase II clinical trials and approval is expected soon. This "haute couture"—custom-made—treatment is relatively complex to produce, test and validate.

This landmark work confirms CureVac's findings that the goal of mRNA vaccines is to boost the immune system in order to thwart the development of cancer cells (by specifically killing them). Cancer vaccines, based on mRNA *transcribed* in vitro and then re-injected into patients, have successfully induced immune responses against some forms of cancer, but, until now, without sufficient control. What is now being done in clinical trials is to combine mRNA-based vaccines with other treatments, such as surgery, immunotherapy, radiotherapy and chemotherapy.

References

1 Jean-Philippe Carralot, Benjamin Weide, Oliver Schoor, Jochen Probst, Birgit Scheel, Regina Teufel, Ingmar Hoerr, Claus Garbe, Hans-Georg Rammensee, Steve Pascolo, "*Production and characterization of amplified tumor-derived cRNA libraries to be used as vaccines against metastatic melanomas,*" Genetic Vaccines and Therapy, August 22, 2005. DOI: https://doi.org/10.1186/1479-0556-3-6.

2 Benjamin Weide, Jean-Philippe Carralot, Anne Reese, Birgit Scheel, Thomas Kurt Eigentler, Ingmar Hoerr, Hans-Georg Rammensee, Claus Garbe, Steve Pascolo, "*Results of the first phase I/II clinical vaccination trial with direct injection of mRNA,*", Journal of Immunotherapy, February- March 2008, vol. 31(2), p. 180–8. DOI: https://doi.org/10.1097/CJI.0b013e31815ce501.

3 Martin Sebastian, Andreas Schröder, Birgit Scheel, Henoch S Hong, Anke Muth, Lotta von Boehmer, Alfred Zippelius, Frank Mayer, Martin Reck, Djordje Atanackovic, Michael Thomas, Folker Schneller, Jan Stöhlmacher, Helga Bernhard, Andreas Gröschel, Thomas Lander, Jochen Probst, Tanja Strack, Volker Wiegand Ulrike Gnad-Vogt, Karl-Josef Kallen, Ingmar Hoerr, Florian von der Muelbe, Mariola Fotin-Mleczek, Alexander Knuth, Sven D Koch, "*A phase I/IIa study of the mRNA-based cancer immunotherapy CV9201 in patients with stage IIIB/IV non-small cell lung cancer,*" Cancer immunology, immunotherapy, May 2019, vol. 68(5), p. 799–812. DOI: https://doi.org/10.1007/s00262-019-023 15-x.

4 Benjamin Weide, Claus Garbe, Hans-Georg Rammensee, Steve Pascolo, "*Plasmid DNA- and messenger RNA-based anti-cancer vaccination,*" Immunology Letters, January 15, 2008, vol. 115(1), p. 33–42. DOI: https://doi.org/10.1016/j.imlet. 2007.09.012.

5 Ugur Sahin, Tim Beissert, Marco Poleganov, Stephanie Herz, *Method for cellular RNA expression,* German patent (BioNTech *AG, TRON,* Johannes Gutenberg University Mainz), No. WO2012072096, international filing date: 3 December

2010, publication date: 5 June 2012. Available at: https://patentscope.wipo.int/search/fr/detail.jsf?docId=WO2012072096&_cid=P10-KML4MS-00900-1.

6 Ugur Sahin, Sebastian Kreiter, Mustafa Diken, Jan Diekmann, Michael Koslowski, Cedrik Britten, John Castle, Martin Löwer, Bernhard Renard, Tana Omokoko, Johannes Hendrikus De Graaf, *Individualized Vaccines for Cancer*, German Patent (BioNTech *AG*, *TRON*, Johannes Gutenberg University Mainz), No. WO2012159643, International Filing Date: 24 May 2011, Publication Date: 29 November 2012. Available at: https://patentscope.wipo.int/search/fr/detail.jsf?docId=WO2012159643&_cid=P10-KML4SI-02597-1.

7 Ugur Sahin, Heinrich Haas, Sebastian Kreiter, Mustafa Diken, Daniel Fritz, Martin Meng, Lena Mareen Kranz, Kerstin Reuter, *RNA formulations for immunotherapy*, German patent (BioNTech, TRON, Johannes Gutenberg University Mainz), no. WO2013143683, international filing date: 25 March 2013, publication date: 3 October 2013. Available at: https://patentscope.wipo.int/search/fr/detail.jsf?docId=WO2013143683&_cid=P10-KML4Y9-04128-2.

8 Lena M. Kranz, Mustafa Diken, Heinrich Haas, Sebastian Kreiter, Carmen Loquai, Kerstin C. Reuter, Martin Meng, Daniel Fritz, Fulvia Vascotto, Hossam Hefesha, Christian Grunwitz, Mathias Vormehr, Yves Hüsemann, Abderraouf Selmi, Andreas N. Kuhn, Janina Buck, Evelyna Derhovanessian, Richard Rae, Sebastian Attig, Jan Diekmann, Robert A. Jabulowsky, Sandra Heesch, Jessica Hassel, Peter Langguth, Stephan Grabbe, Christoph Huber, Özlem Türeci, Ugur Sahin, "*Systemic RNA delivery to dendritic cells exploits antiviral defense for cancer immunotherapy,*", *Nature*, 16 June 2016, vol. 534(7607), pp. 396–401. DOI: https://doi.org/10.1038/nature18300.

9 Ugur Sahin, Evelyna Derhovanessian, Matthias Miller, Björn-Philipp Kloke, Petra Simon, Martin Löwer, Valesca Bukur, Arbel D. Tadmor, Ulrich Luxemburger, Barbara Schrörs, Tana Omokoko, Mathias Vormehr, Christian Albrecht, Anna Paruzynski, Andreas N Kuhn, Janina Buck, Sandra Heesch, Katharina H. Schreeb, Felicitas Müller, Inga Ortseifer, Isabel Vogler, Eva Godehardt, Sebastian Attig, Richard Rae, Andrea Breitkreuz, Claudia Tolliver, Martin Suchan, Goran Martic, Alexander Hohberger, Patrick Sorn, Jan Diekmann, Janko Ciesla, Olga Waksmann, Alexandra-Kemmer Brück, Meike Witt, Martina Zillgen, Andree Rothermel, Barbara Kasemann David Langer, Stefanie Bolte, Mustafa Diken, Sebastian Kreiter, Romina Nemecek, Christoffer Gebhardt, Stephan Grabbe, Christoph Höller, Jochen Utikal, Christoph Huber, Carmen Loquai, Özlem Türeci, "*Personalized RNA mutanoma vaccines mobilize poly-specific therapeutic immunity against cancer,*" *Nature,* July 13, 2017, vol. 547(7662), p. 222–226. DOI: https://doi.org/10.1038/nature23003.

14

Experiments and Clinical Trials Carried Out in Other Therapeutic Fields

- **Replacement proteins**

The production of mRNA-generated replacement proteins today represents an enormous technical challenge for the treatment of various diseases. Indeed, the translation process (described in Part I, Chap. 3, "Messenger RNA, from Transcription to Protein Translation") does not result in the synthesis of a fully and immediately functional protein. The polypeptide, once translated, undergoes a series of modifications in distinct cellular structures that complete the functionalization of the protein. Proteins can undergo several modifications. The most common is glycosylation, which adds a carbohydrate moiety to the already formed protein, and which mainly concerns membrane proteins. There is also phosphorylation, which consists in adding a phosphate group to an already formed protein, and which plays an essential role in intracellular communication.

The following are the most significant milestones in these alternative protein experiments:

1992—Jirikowski demonstrates for the first time that it is possible to produce replacement proteins using mRNA: rats express the protein vasopressin (or antidiuretic hormone) after injection of mRNA coding for this protein into their hypothalamus [1]. The study found that the injection was temporarily effective against diabetes insipidus in Brattleboro rats, a type of diabetes characterized by very intense thirst and the production of large quantities of very dilute urine;

© The Author(s), under exclusive license to Springer Nature
Switzerland AG 2023
J. Lemonnier and N. Lemonnier, *The Marathon of the Messenger*,
https://doi.org/10.1007/978-3-031-39300-6_14

Brattleboro rats are a specific strain of laboratory rats, bred in 1961, in West Brattleboro, Vermont, USA. Due to a natural genetic mutation, these rats could not synthesize the hormone vasopressin, and therefore suffered from diabetes insipidus.

2005–2010—Drew Weissman and Katalin Kariko devote most of their work to the development of replacement proteins from modified mRNA, as previously reported. It should be recalled that patent US8278036B2, filed in 2005: "RNA containing modified nucleosides and related methods of use", of which they are the inventors (cf. *supra* Part II, Chap. 10, "Modified mRNA and Unmodified mRNA: For What Purpose?"), protects "*RNA, oligoribonucleotide and polyribonucleotide molecules comprising pseudo-uridine or a modified nucleoside, gene therapy vectors comprising the aforementioned compounds, processes for synthesizing these compounds, as well as processes for gene replacement, gene therapy, silencing[1] of gene transcription, and delivery of therapeutic proteins to tissues in vivo, comprising the molecules*" [2].

The goal of this mRNA "de-immunization" is to translate proteins that are fully functional and free of inflammatory response, which is, as we have seen, the opposite of vaccination. In fact, Weissman and Kariko continued to pursue this goal until at least 2017.

These two researchers are working, in particular, on the production of erythropoietin (EPO) from modified mRNA (via the replacement of the uracil by pseudo-uridine in order to be invisible to the immune system). The objective is to treat anemia and certain respiratory diseases: EPO, a hormone-like protein, increases the number of red blood cells in the blood, and thus promotes better oxygenation of the body [3]. Preclinical trials were conducted on mice and monkeys in the 2010s.

At present, however, there are still major obstacles: the high cost of the method, which involves, in particular, expensive purification of the mRNA transcribed in vitro, the sometimes narrow therapeutic windows, and problems of access to certain organs of the human body. Therefore, researchers are working on proteins expressed in easily accessible organs, such as the liver. Despite these difficulties, therapies based on the production of replacement proteins from mRNA represent a promising avenue with a large number of possible clinical applications.

Of particular note is Moderna's current Phase II clinical trial for the treatment of myocardial infarction. Once administered to the patient, the mRNA

---

[1] **Silencing** is a genetic manipulation technique that aims to reduce or prevent the expression of a protein from its corresponding gene.

will express the VEGF (*Vascular Endothelial Growth Factor*) protein for a few days. This then triggers the formation of new blood vessels in the heart around the clot. Phase I of this clinical trial was conducted by skin injection—only to verify that this mRNA causes new blood vessels to form and that it is safe—while Phase II is currently being conducted by injecting the mRNA directly into the ischemic heart during the transient cessation of blood flow during the surgical procedure.

At the European regulatory level, the use of mRNA transcribed *in vitro* for the synthesis of replacement proteins is now considered to be gene therapy!

Indeed, the directive 2009/210/EC of September 14, 2009, of the European Medicines *Agency* (EMA) defines a gene therapy medicinal product as: "*a biological medicinal product which has the following characteristics: (a) it contains an active substance which contains or consists of a recombinant nucleic acid administered to persons with a view to regulating, repairing, replacing, adding or deleting a genetic sequence (nucleic acid = DNA or RNA); (b) its therapeutic, prophylactic or diagnostic effect is directly dependent on the recombinant nucleic acid sequence it contains or on the product of the genetic expression of that sequence*".

While mRNA vaccines against infectious diseases are not considered gene therapy drugs, replacement proteins produced with mRNA should logically be considered gene therapy drugs if they are marketed after clinical trials in humans. The pre-marketing controls for drugs that use mRNA would therefore be more demanding and complex for replacement proteins than for vaccines! Surprisingly, mRNA vaccines are often categorized as "gene therapy", because they are produced by a biological process—transcription—even though they do not alter the genome.

● **Genetic diseases**

Phase I clinical trials of mRNA therapy are underway for certain genetic diseases that result in the absence of a particular protein. Examples include:

● cystic fibrosis, a well-known genetic disease that results, in particular, in lung damage: mRNA therapy aims to revitalize the lungs through long-term treatment (Translate Bio biotech). The most recent clinical trial consisted in the subjects inhaling the mRNA encoding the CFTR gene, which is translated into a functional protein by the lung cells; this trial, however, proved to be a failure.

Cystic fibrosis is linked to mutations in the CFTR gene on chromosome 7, resulting in an alteration of the CFTR (*cystic fibrosis transmembrane conductance regulator*) protein. The dysfunction of this protein causes an increase in the viscosity of mucus and its accumulation in the respiratory and digestive tracts. The disease affects many organs, but respiratory damage is predominant and accounts for most of the deleterious health consequences. The most frequent clinical form combines respiratory disorders, digestive disorders and growth disorders.

- *Surfactant Protein B Deficiency*: Ethris has developed a spray to treat this other genetic disease, which also affects the lungs.

Surfactant Protein B Deficiency is a very rare genetic disease (one in a million) characterized by the deficiency of the surfactant protein B. Surfactant is a complex material that is essential for normal respiratory function. Respiratory distress is visible from birth. The disease causes very severe damage to the lungs, requiring mechanical ventilation and a heart bypass. A lung transplant is the only lasting treatment. Life expectancy is very short.

In both cases, the in vitro transcribed mRNA contained in the spray is delivered through the bronchial tree, where it transfects lung cells. The very short half-life of the mRNA requires chronic treatment and continuous immune monitoring.

Another very interesting application of mRNA therapy to treat genetic diseases is the CRISPR-Cas9 technology, known as "genetic scissors". Discovered by Emmanuelle Charpentier and Jennifer Doudna, both Nobel Prize in Chemistry 2020, Cas9 is a bacterial protein that specifically cuts a sequence in the genome. The concept is to use mRNA to produce this Cas9 protein, which will then modify the targeted DNA sequence. As stated in *Campbell Biology*, this technique is very effective in inactivating a given gene in order to study its function. Researchers have also modified it in order to use the CRISPR-Cas9 complex to repair a mutated gene. They introduce a segment from the normal (functional) gene along with the CRISPR-Cas9 system. Once the target DNA is cut by the Cas9 protein, the repair enzymes can use a segment of normal DNA as a template to repair the targeted DNA at the break point [4]. Gene therapy can therefore utilize CRISPR-Cas9 using Cas9-encoding mRNA. The first demonstration that this technique can cure humans with an inherited disease was published in 2021 [5].

mRNA can also be used to treat genetic skin diseases and, by extension, to regenerate and rejuvenate the skin. For example, the protein elastin is produced until the end of adolescence, after which it is hardly ever secreted. This protein has elasticity properties, which explain, in particular, why the skin regains its initial shape after a pinch or a stretch. Elastin remains in the dermis for years and loses its mechanical properties over time. By injecting mRNA encoding elastin into the skin, it is thus possible to restore elasticity, which will then persist, because this protein, produced for a few hours following the injection, will persist in the dermis for years [6].

- **Regenerative medicine**

In regard to repairing or replacing damaged tissues or organs, mRNA also has its place in regenerative medicine, as it can "reprogram" adult cells from one type to another, through a process of transdifferentiation.

Transdifferentiation is defined by the fact that a somatic cell loses its normal characters and acquires new characters or functions: in fact, each cell can transform into another cell type if it receives the right external signal. The mRNA can be the carrier of this signal, considering its determining role in cellular identity.

In this regard, as Ana del Pozo-Rodriguez's team states in the article *"Nanomedicines to Deliver mRNA: State of the Art and Future Perspectives"*, published in the journal *Nanomaterials*, in 2020 [7]: *"Transdifferentiation of somatic cells with IVT mRNA has been mainly focused on the generation of insulin secreting β-cells for type 1 diabetes patients and on obtaining cardiomyocytes to regenerate the cardiac tissue after a heart attack. mRNA is a promising tool for the delivery of factors targeting altered signaling pathways in the early hours of infarction, as well as to address experimental and clinical needs to regenerate cardiac tissue and cardiac function in ischemic heart disease"*.

This article demonstrates the interest in mRNA for the regeneration of bone, connective tissue and cardiac tissue (in case of heart attack or ischemia). It also reports a preclinical study on mice, which demonstrates the possibility of reprogramming cardiac fibroblasts (cardiac muscle support cells) into cardiomyocites (cardiac contractile muscle cells) by mRNA transfection [8].

Although still in preclinical studies, the generation of β-cells or cardiomyocytes would open up a prime therapeutic avenue for mRNA in regenerative medicine, harnessing mRNAs to boost regeneration of severely damaged tissues.

● **Allergies**

The interest in mRNA to better fight against allergies and hypersensitivity phenomena, caused by an excessive immune response of the body against certain antigens called allergens (for example, pollens), is also significant.

In concrete terms, certain antibodies in the body attach themselves specifically to allergens and bind to mast cells, the connective tissue cells responsible for the inflammatory reaction. The mast cells then release an excess of substances, such as histamines, and the allergic reaction is triggered with a host of specific symptoms: redness, sneezing, tears, possible breathing difficulties, etc.

By encoding the allergen, the mRNA will trigger a prophylactic immune response. The individual will thus be protected in case of exposure to the natural allergen. The immune response is that of mRNA vaccination against infectious diseases: activation of helper T lymphocytes, followed by production of antibodies that confer immunity on the patient.

Richard Weiss, Sandra Scheiblhofer, Elisabeth Roesler, Fatima Ferreira, and Josef Thalhamer demonstrated the value of mRNA vaccination against allergies as early as 2010 [9, 10].

● **Certain hormonal dysfunctions**

The possibility of producing antidiuretic hormone (or vasopressin), a peptide hormone, was demonstrated in 1992 by Jirikowski et al. [1]. Moreover, as insufficient insulin production by the pancreas is the cause of type 2 diabetes, it would be possible to use an mRNA encoding insulin in the treatment of this type of diabetes.

Hormone receptors can also be considered: some rare genetic diseases can be linked to such a dysfunction, such as Laron syndrome. People with this disease are very short; they also suffer from delayed motor development and delayed puberty. This congenital condition is linked to a mutation in the growth hormone receptor (GHR) gene. Mutations in the extracellular domain of the receptor cause a decrease in growth hormone levels.

These three examples illustrate the therapeutic potential of mRNA in the face of under-expression or deficiency of hormones, or even receptors. And there are many pathologies, of somatic or genetic origin, that can benefit from mRNA therapy!

References

1 Jirikowski GF, PP Sanna, D. Maciejewski-Lenoir, FE Bloom, *"Reversal of diabetes insipidus in Brattleboro rats: intrahypothalamic injections of vasopressing mRNA"*, Science, February 21, 1992, 255 (5047), 996–998. DOI: https://doi.org/10.1126/science.1546298.

2 Katalin Kariko, Drew Weissman, *"RNA containing modified nucleosides and methods of use thereof"*, U.S. Patent (University of Pennsylvania), N°. US8278036B2, publication date: August 23, 2005. Available at: https://patents.google.com/patent/US8278036B2/en.

3 K. Kariko, Hiromi Muramatsu, Jason M. Keller, Drew Weissman, *"Increased erythropoiesis in mice injected with submicrogram quantities of pseudouridine-containing mRNA encoding erythropoietin"*, Molecular Therapy Journals, May 2012, vol. 20 (5), p. 948–953. DOI: https://doi.org/10.1038/mt.2012.7.

4 *Campbell Biology*, 11th edition by Lisa Urry, Michael Cain, Steven Wasserman, Peter Minorsky, Jane Reece, Pearson Education, 2017.

5 Julian D Gillmore, Ed Gane, Jorg Taubel, Justin Kao, Marianna Fontana, Michael L. Maitland, Jessica Seitzer, Daniel O'Connell, Kathryn R. Walsh, Kristy Wood, Jonathan Phillips, Yuanxin Xu, Adam Amaral, Adam P. Boyd, Jeffrey E. Cehelsky, Mark D. McKee, Andrew Schiermeier, Olivier Harari, Andrew Murphy, Christos A Kyratsous, Brian Zambrowicz, Randy Soltys, David E. Gutstein, John Leonard, Laura Sepp-Lorenzino, David Lebwohl, *"CRISPR-Cas9 In Vivo Gene Editing for Transthyretin Amyloidosis"*, New England Journal of Medicine, Aug. 5, 2021, vol. 385 (6), pp. 493–502. DOI: https://doi.org/10.1056/NEJMoa2107454.

6 Lescan M, Perl RM, Golombek S, Pilz M, Hann L, Yasmin M, Behring A, Keller T, Nolte A, Gruhn F, Kochba E, Levin Y, Schlensak C, Wendel HP, Avci-Adali M, *"De Novo Synthesis of Elastin by Exogenous Delivery of Synthetic Modified mRNA into Skin and Elastin-Deficient Cells"*, Molecular therapy. Nucleic Acids, March 30, 2018, vol. 11; p. 475–484. DOI: https://doi.org/10.1016/j.omtn.2018.03.013.

7 Itziar Gómez-Aguado, Julen Rodríguez-Castejón, Mónica Vicente-Pascual, Alicia Rodríguez-Gascón, María Ángeles Solinís, Ana del Pozo-Rodríguez, *"Nanomedicines to Deliver mRNA: State of the Art and Future Perspectives"*, Nanomaterials, 20 February 2020. DOI: https://doi.org/10.3390/nano10020364.

8 Kunwoo Lee, Pengzhi Yu, Nithya Lingampalli, Hyun Jin Kim, Richard Tang and Niren Murthy, *"Peptide-enhanced mRNA transfection in cultured mouse cardiac fibroblasts and direct reprogramming toward cardiomyocyte-like cells"*, International Journal of Nanomedicines, March 6, 2015. DOI: https://doi.org/10.2147/IJN.S75124.

9 Richard Weiss, Sandra Scheiblhofer, Elisabeth Roesler, Fatima Ferreira, Josef Thalhamer, "*Prophylactic mRNA vaccination against allergy*", *Current Opinion in Allergy and Clinical Immunology*, December 2010, vol. 10(6), p. 567–74. DOI: https://doi.org/10.1097/ACI.0b013e32833fd5b6.

10 Sandra Scheiblhofer, Josef Thalhamer, Richard Weiss, "*DNA and mRNA vaccination against allergies*", *Pediatric Allergy and Immunology*, November 2018, vol. 29(7); p. 679–688. DOI: https://doi.org/10.1111/pai.12964.

15

Modified mRNA *Versus* Unmodified mRNA: Not just a Scientific Issue

- **What conclusions can be drawn from the studies and clinical trials?**

To conclude this section on seminal studies and initial preclinical and clinical experiments with mRNA, a comparative analysis of the use of pseudouridine-modified synthetic mRNA and unmodified synthetic mRNA can now be made in the form of the following table (Table 15.1).

To the question "Are there specific arguments that justify the choice of using modified, rather than unmodified, mRNA for vaccination?", it appears that there is no obvious answer. Indeed, behind this question, another one arises: **is it possible to induce a good immune response (antibodies and T lymphocytes) by using unmodified, as well as modified, mRNA?** On this precise subject, we have reported above on convincing studies carried out using either modified or unmodified mRNA. It is clear that there is no definitive answer in either direction, and that the current scientific debate on the subject is important…

In their excellent 2019 paper [1], Verbeke et al. thus frame the terms of the modified mRNA/unmodified mRNA debate in a particularly apt way when they describe two possible strategies:

(i) A strategy that consists of using an mRNA whose formulation will be designed and perfected to obtain a strong adjuvant effect,

(ii) A strategy that consists of *"working with a 'modified' mRNA with a high translation capacity, and thus a better bioavailability of the antigen (…).*

© The Author(s), under exclusive license to Springer Nature Switzerland AG 2023
J. Lemonnier and N. Lemonnier, *The Marathon of the Messenger*,
https://doi.org/10.1007/978-3-031-39300-6_15

Table 15.1 Comparative interests of modified and unmodified mRNA

Type of mRNA	Main uses	Objective given the characteristics of mRNA	Potential immune action	Existence of an inflammatory reaction	Efficiency
mRNA modified by PseudoU	Protein alternatives	Producing a therapeutic protein	Not desired	Inflammatory reaction avoided: no reaction of mRNA by PPR receptors	+++
Unmodified mRNA	Vaccines	Expression of a tumor (cancer) or exogenous (infectious diseases) antigen	Desired	Induced immune response: production of interferon-1, which blocks protein synthesis, except in certain specialized cells, and activates adaptative immunity	+++

Verbeke goes on to say that *"with regard to vaccination, it must be taken into account that modifications that promote the translational capacity of mRNA involve a partial or complete reduction of the interaction between mRNA molecules and one or more PRRs capable of detecting a particular virus (…). As such, this could be at the expense of the adjuvant effect of the mRNA vaccine. After all, the two outcomes are regulated in opposite ways by type 1 IFN-induced genes. Priority is often given to the translational capacity of mRNA with the idea of improving antigen availability. Yet, from an immunological perspective, mRNA detection by innate immunity, which evokes phenotypic immune profiling and the cytokine milieu, is at least as important. Indeed, these innate immune signals will trigger and guide the selection of effector responses, which is of critical importance for the therapeutic value of the vaccine"*.

More recently, there was a surprising report in a paper of a type 1 interferon response after injection of BioNTech's mRNA-modified BNT162b2 vaccine, induced by a signaling pathway other than TLR 7—the MDA5 pathway [2]. This demonstrates the close links, still poorly understood today, between innate and adaptive immunity.

Finally, we mentioned above the clinical success of a new generation of mRNA vaccines, based on a self-replicating mRNA that, from a limited dose of unmodified mRNA, will make possible a much faster translation of viral antigens through a replicase enzyme.

This seems to us to demonstrate that, for the development of mRNA vaccines, unmodified mRNA can be used just as well as pseudo-uridine-modified mRNA. Moreover, pseudo-uridine-modified mRNA is not the most widely used type of mRNA at the present time: CureVac and BioNTech are using mainly unmodified mRNA to develop and test cancer vaccines.

- **Modified mRNA selected by BioNTech for Covid vaccination**

It is thus interesting to study the reasons why BioNTech finally retained the modified mRNA for the Covid vaccination.

Recall that, in 2017, BioNTech obtained a sublicense to exploit the University of Pennsylvania's "pseudo-uridine-modified mRNA" patent. Yet, in the joint BioNTech/Cellscript press release for that sublicense agreement, it was made clear that the German biotech would continue to use its other unmodified mRNA-based technologies for the development of personalized immunotherapies: *"The modified mRNA technology is complementary to, but separate from, BioNTech's other mRNA technologies used in its clinical-stage cancer vaccines for the treatment of cancer"*. [3]

However, at present, in the fight against Covid-19, the only mRNA vaccines already approved, those from BioNTech and Moderna, are modified mRNA. Moderna has only developed a modified mRNA vaccine candidate, while BioNTech has tested both modified (BNT162b1 and BNT162b2) and unmodified (BNT162a1) mRNA vaccine candidates.

For Covid vaccination, Moderna uses only the modified mRNA and does not compare it with the unmodified mRNA. CureVac uses only unmodified mRNA and does not compare it with modified mRNA. BioNTech, on the other hand, uses unmodified mRNA for vaccines and modified mRNA for protein replacements. However, in 2020, BioNTech tested both mRNAs (modified and unmodified) in its Phase I trials and then decided, with Pfizer, to use the modified mRNA for subsequent phases without justifying its choice with clinical results (results obtained with the unmodified mRNA are not published).

Although not formally stated, the reason for BioNTech's choice to retain the modified mRNA is clearly suggested in Joe Miller's book, written with Özlem Türeci and Ugur Sahin, *The Vaccine. Inside the Race to Conquer the Covid-19 Pandemic* [4].

At the time of the development of the anti-covid vaccines, BioNTech (and also Moderna) had chosen an intramuscular injection, and they had LNP with additional immunostimulating properties when used with this mode of administration. This shows that the adjuvant effect obtained is the consequence of a complicated parameterization of the formulation, namely, the result of the subtle interaction amongthe container (the lipid nanoparticle), the mode of administration (intravenous, intramuscular, subcutaneous) and the content (the nucleic acid, in this case, mRNA, integrated into the liposome). As a consequence, mRNA modification appears to be an adjustment variable that may or may not be useful, depending on formulation, mRNA sequence, contaminants, dose and injection site.

Therefore, the liposomes available to BioNTech in spring 2020 already had an adjuvant effect, so there was concern that an excessive immune response might be elicited when using unmodified mRNA. The BioNTech team could then have subjected the unmodified mRNA to a specific purification process, but, indeed, it would have taken a great deal of time, and Ugur Sahin did not want to add an additional level of complexity. The clinical trial therefore began with an unpurified form of unmodified mRNA.

The fear of an excessive immune response was justified in the preclinical trial with 100 μg of unmodified mRNA: the injected rats had a very strong inflammatory reaction. This led to the dilution of the unmodified mRNA vaccine that was administered to patients in the clinical trials; it is very likely that this dilution led to unsatisfactory results, eliciting a weak antibody response and poor protection against Sars-Cov-2.

The choice of BioNTech's modified mRNA for the anti-covid vaccine was therefore justified by clinical and practical reasons: the modified mRNA was a natural choice given the liposomes available to BioNTech for intramuscular injections and the results observed during preclinical and clinical trials. And the urgency of the health crisis made it imperative to make quick decisions. The rapid development of an effective vaccine was rightly seen as the most effective way to address the pandemic.

This choice of modified mRNA had other consequences, which we now need to address.

- **Modified mRNA and Covid vaccination: a high point for Drew Weissman, Katalin Kariko… and the U.S.!**

Recall that patent US8278036B2, which protects the modified mRNA synthesis technology that uses the Kariko/Weissman invention, is currently held by the University of Pennsylvania (or UPENN) in the United States, and was exclusively licensed to a Wisconsin-based U.S. startup, Cellscript, in 2016.

The research that led to this innovation was partially funded by the *National* Institutes of Health, and it is very likely that the U.S. government also receives royalties on this patent [5]. In the description of the patent, the *Statement Regarding Federally Sponsored Research Or Development* states: "*The invention described herein was supported in part by grants from The National Institutes of Health (Grant No. AI060505, AI50484 and DE14825). The U.S. Government may have certain rights in this invention*". Indeed, the licenses to the patent include a "U.S. Government Rights" clause, which states that they are "*expressly subject to all applicable U.S. Government rights, including, but not limited to, any applicable requirement that the products resulting from this intellectual property and sold in the United States be substantially manufactured in the United States…*".

This is recognition that these licenses are *ultimately* subject to the *Bayh Dole Act*. According to this 1980 law, American universities can patent their discoveries and, therefore, benefit from substantial financial rewards by granting licenses to exploit them; in return, the American government receives an irrevocable and non-transferable license fee, while the patent holder (the University of Pennsylvania) must expressly favor American companies.

The sub-licenses obtained by BioNTech and Moderna to exploit the patent, which were granted at a high price, are therefore of great benefit to the U.S. Important players in the USA are clearly benefiting: the U.S. company Cellscript, the University of Pennsylvania and the U.S. Government.[1] Clearly, behind the health aspects of modified mRNA, there are real issues of intellectual property—and of American political influence!

For research laboratories and biotechs located in Europe, it appears, under these conditions, a priori preferable to develop therapies based on unmodified mRNA.

The attempted takeover of CureVac by the U.S. government in March 2020, which will be discussed later, takes on a very special significance in this context.[2]

The two inventors of the University of Pennsylvania patent, Katalin Kariko and Drew Weissman, received millions of euros, considering the very advantageous terms of remuneration for patent inventors defined by the University of Pennsylvania: their share amounts to 30% of the revenue received.

> This is a result of the *Bayh Dole Act,* which requires agencies receiving federal grants to share royalties from patent licenses with inventors. See "Patent and Tangible Research Property Policies and Procedures," available at https://catalog.upenn.edu/pennbook/patent-tangible-research-property-policies-procedures/.

In addition, many media outlets, as of December 2020, are touting them as Nobelizable, in that the pseudouridine-modified mRNA they originated is now being used worldwide for BioNTech and Moderna mRNA vaccines. Finally, after having received many valuable awards, especially the famous Breakthrough Prize in 2022, Katalin Kariko and Drew Weissman receive on 2 October 2023 the Nobel Prize in Physiology and Medicine "for their discoveries concerning nucleoside base modifications that enabled the development of effective mRNA vaccines against Covid-19", as indicated by the press release of the Nobel Committee.

Perhaps it is the extraordinary emphasis on his invention that has led Drew Weissman to repeatedly assert the essential nature of the mRNA modification? As, for example, in this interview given at the beginning of 2021 to the

---

[1] Intellectual property issues are of critical importance. See *infra* Part III, Chap. 18, "The Importance of Intellectual Property and Patents".

[2] Given CureVac's expertise in mRNA vaccines, who would have thought that their first phase-III clinical trial of a Covid vaccine would fail so miserably?

American magazine *MIT Technology Review* [6], in which he states: "*The first attempt to use synthetic messenger RNA to make an animal produce a protein was in 1990. It worked but a big problem soon arose. Their fur gets ruffled. They lose weight, stop running around. Give them a large dose, and they'd die within hours. We quickly realized that messenger RNA was not usable. The culprit was inflammation. (...)*". As we have seen, since the mRNA injections performed as early as 1993 by Pierre Meulien and Frédéric Martinon, the inflammatory reaction has always been controlled. And Weissman is not unaware of the work done over the last twenty years by CureVac and BioNTech! (see Part II, Chap. 12 *above*, "Experiments and Clinical Trials Against Infectious Diseases").

These considerations should not, of course, lead one to underestimate the importance of the development of pseudo-uridine-modified mRNA for treating a wide variety of diseases, primarily in the context of a "protein replacement" approach and now also in vaccinology.

Nor should they lead one to underestimate the role played by Katalin Kariko. It is certain that her media aura, skilfully maintained by the University of Pennsylvania, has contributed in an indisputable way to the current reputation of therapeutic mRNA technologies. Otherwise, it would certainly have taken many more years to recognize mRNA-based therapies as very promising avenues of medical research—which they are today, in the eyes of all.

And indeed, it took a long time—more than 20 years—before mRNA vaccination came into the public eye with a bang, in the year 2020.

References

1 Rein Verbeke, Stefaan De Smedt, Ine Lentacker, Heleen Dewitte, "*Three decades of messenger RNA vaccine development*", in *Nanotoday*, October 2019, vol. 26. DOI: https://doi.org/10.1016/j.nantod.2019.100766.

2 Chunfeng Li, Audrey Lee, Lilit Grigoryan, Prabhu S Arunachalam, Madeleine K D Scott, Meera Trisal, Florian Wimmers, Mrinmoy Sanyal, Payton A Weidenbacher, Yupeng Feng, Julia Z Adamska, Erika Valore, Yanli Wang, Rohit Verma, Noah Reis, Diane Dunham, Ruth O'Hara, Helen Park, Wei Luo, Alexander D Gitlin, Peter Kim, Purvesh Khatri, Kari C Nadeau, Bali Pulendran, "*Mechanisms of innate and adaptive immunity to the Pfizer-BioNTech BNT162b2 vaccine*", *Nature Immunology*. 2022 Apr; 23(4):543-555. Doi: https://doi.org/10.1038/s41590-022-01163-9. Epub 2022 Mar 14.

3 "BioNTech AG enters into a license agreement with CELLSCRIPT, LLC as it advances the development of bispecific antibodies encoding messenger RNA

and other therapeutic proteins", press release available on the internet at https://www.globenewswire.com/news-release/2017/09/06/1108043/0/en/BioNTech-AG-Enters-into-Licensing-Agreement-with-CELLSCRIPT-LLC-as-it-Advances-Development-of-Messenger-RNA-Encoding-Bispecific-Antibodies-and-other-Therapeutic-Proteins.html.

4 Joe Miller, with Özlem Türeci and Ugur Sahin, "*The Vaccine: Inside the Race to Conquer the Covid-19 Pandemic*", February 2022, St. Martin's Press.

5 Katalin Kariko, Drew Weissman, "*RNA containing modified nucleosides and methods of use thereof*", U.S. Patent (University of Pennsylvania), No. US8278036B2, publication date: August 23, 2005. Available at: https://patents.google.com/patent/US8278036B2/en.

6 Antonio Regalado, "*The next act for messenger RNA could be bigger than covid vaccines*", *MIT Technology* Review, February 5, 2021. Article available at: https://www.technologyreview.com/2021/02/05/1017366/messenger-rna-vaccines-covid-hiv/.

Part III

The Triumph of Messenger RNA

16

A Very Progressive Diffusion in Scientific Circles

Until the mid-2010s, the mRNA vaccination approach had very limited diffusion. Despite the promising results obtained by CureVac and BioNTech, the scientific community remained hesitant, or even outright reluctant, for a long time.

Sahin, Kariko, and Türeci's 2014 review in *Nature* [1] marks an important step toward broader recognition of mRNA therapies, but it does not signal the end of skepticism. However, the article timely points to the decisive contribution of two decades of work on mRNA therapies in all areas of medicine: vaccination and *protein replacement*.

Why has the reality of scientifically established facts not been disseminated more rapidly? And this, in a world of communication with exponential development since the advent of micro-computing in the 1980s.

The answer to this question is based on several factors.

a. **The incredulity of minds in the face of any new concept and the single thought prevailing in too many laboratories**

Today, it is widely accepted that scientific consensus is a good thing. But this is not necessarily the case. To demonstrate (or not!) the potential of a technology is almost always a lifetime's work. After promising initial studies, it is often necessary to win over the conviction of one or more influential researchers. Then, often in a very progressive way, to obtain the adhesion and the recognition of all. Finally, everyone forgets that there have always

© The Author(s), under exclusive license to Springer Nature
Switzerland AG 2023
J. Lemonnier and N. Lemonnier, *The Marathon of the Messenger*,
https://doi.org/10.1007/978-3-031-39300-6_16

been divisions within the scientific community, even in regard to work that is universally recognized today, such as that of Louis Pasteur.

Just as in other sciences, some theories, at their beginning, have been decried by the majority of their adherents' peers. Conformism and/or the difficulty of accepting an innovative conception that challenges certain religious or ideological beliefs explain this.

> Let's remember the Austrian doctor Semmelweis, who, in the 1850s and 1860s, recommended that caregivers wash their hands thoroughly before giving birth! The lack of hygiene of the surgeons was, at the time, responsible for the death, by puerperal fever, of thousands of young women giving birth. Despite the spectacular results, Semmelweis was disowned by his peers and banished from the medical profession.

The very first human mRNA vaccination trials, conducted at the University Hospital of Tübingen in collaboration with CureVac, could not find publication in major medical journals such as *Nature, Science* or *The Lancet*. They were only accepted by journals with a lower impact. For example, Steve Pascolo's seminal paper [2] appeared, in 2004, in *Expert Opinion on Biological Therapy*. Thereafter, the ambient interest of the scientific and medical community has been very gradual. This is due, in particular, to the growing appeal of mRNA therapeutics to American start-ups and their investors.

This relative conservatism of the scientific community is also favored by the fact that the peer *reviewers* of a journal not only ensure the accuracy of the scientific elements provided and the coherence of the elements contributed in any new article, but also verify the adequacy, in their opinion, between its message and the impact of the journal (cf. *below,* the section on "The importance of the impact factor of the publication").

b. The pre-eminence of Anglo-Saxon scientific and medical journals

While CureVac's initial work was not cited (or rarely so), the American studies, although they came much later, were cited more regularly. It is well known that researchers working in the United States are more likely to report on American research.

The review article by Sahin, Kariko and Türeci in *Nature* in 2014 mentioned several times previously [1] is a good example: the 256 references cited at the end of the article are overwhelmingly those of American researchers; as for the research carried out by CureVac in particular, it is not even mentioned.

There is, in fact, a language bias in favor of English-language journals. The two scientific journals considered the most prestigious are *Nature* (English) and *Science* (American). English, which has become the global language in the scientific and technological fields, actually favors Anglo-Saxon journals. Journals in other languages (French, German, Italian, etc.) have limited distribution in their countries of origin, despite the current very efficient translation systems. These national journals concern articles of specialties or general reviews. Today, all scientific and medical researchers write, communicate and express themselves in English.

The issue is also closely linked to the impact factor issue.

"Today, all scientific or medical researchers write, communicate and express themselves in English."

c. **The importance of the impact factor of the publication**

The *impact factor (IF)* is an essential element nowadays for the diffusion of scientific knowledge.

The impact factor is an indicator that measures the visibility of a scientific journal. For a given year, the IF of a journal is equal to the average of the number of citations of the articles of this journal published during the two previous years. A journal with a high IF would thus be considered more important (because it is more visible: more widely read and more often cited) than a journal with a low IF. A researcher wishing to see their article cited a lot will therefore target a journal with a high impact factor in their scientific discipline.

This notion of impact factor has a certain influence in the field of scientific publication. However, certain imperfections associated with this indicator lead us to put its relevance into perspective. We can thus mention:

- The number of citations is not a correct measure of the quality or even quantity of new information in publications. Indeed, citations are considered according to their number, and not according to their nature. In practice, there are criteria other than the raw scientific quality of the work that cause an article to be read and cited: the clarity of the article, the fact that it deals with a "fashionable" theme, or not, or even that the article is cited as an example of bad scientific research. Therefore, the impact factor is not appropriate to evaluate an individual or to assess the quality of all articles in a journal.
- Moreover, since it is based on the number of citations, the IF of a journal can artificially increase through "self-citation". This term refers to the citation of articles in a journal by other articles in the same journal. The editorial policy of some journals may then encourage them to mention only the authors of works published by these same journals, in order to raise an IF considered too low. This deleterious bias also favors researchers who publish in the most widely read journals, without this necessarily being associated with the quality of the publications. It is even an incentive to cut up results in order to multiply publications!
- **To be visible, the scientist must imperatively publish in an English language journal of high IF. As the promotion of an academic researcher depends on the quality of their work qualified by the IF of their publications, for each one, it is a question of interesting the editorial staff of the highest IF journal, and as soon as possible!**

The researcher is then confronted with certain dilemmas, detailed in the box "Publish high or publish safe?".

Under these conditions, the impact factor appears to be such a significant indicator for researchers that they may "ignore" low impact publications in references that relate to previous research, even if they are perfectly aware of them.

All these elements may explain the "ignorance" of high *impact* journals regarding the first articles published on vaccine mRNA.

Thus, with regard to the first studies carried out in Tübingen by Karl-Eberhard University and by CureVac, which were published in a journal with a moderate IF, one must consider the objective: for this biotech, which was not academic but economic, financed by investors, and, in particular, by *business angels*, obtaining the visa authorizing the clinical trial was the *primum movens*.

> A *business angel* is a natural or legal person who decides to invest part of his or her financial assets in innovative companies with high potential.

Once this difficult stage was over, publication, that is to say, its distribution vector, took second place. It was necessary to set a date, and as quickly as possible. In a context of mistrust of an innovative approach far from the single thought, journals with a high IF could have rejected the publication, or even used delaying tactics to slow it down: so, it seemed safer to address reading committees with a lower IF, with rapid evaluation and revision processes, and who were happy to make a scoop to increase their IF!

Publication in higher IF journals would undoubtedly have helped CureVac to obtain more funding from investors. Indeed, investors, even if they often say the opposite, are often advised by scientists.

In fact, the trigger for CureVac came at the end of 2005: German billionaire Dietmar Hopp, founder of the large German group SAP, invested 35 million euros in the Tübingen biotech. The influential biotech entrepreneur Friedrich von Bohlen und Halbach had introduced him to CureVac's founders a few months earlier…

16.1   Publish High or Publish Safe?

The researcher who has completed their experimental study generally wishes to make the results known, possibly after the corresponding patent has been filed.

They then write their article for the most appropriate Impact Factor (IF) journal, based on the power of the message delivered, its novelty, and the importance of the data produced.

In doing so, they accept the risk of losing precious weeks, which actually correspond to the time needed for the evaluation of the article by the reading committee. The latter can, in fact, reject the submitted article after an indeterminate period. In the latter case, the researcher will "go down" the IF scale in hopes of finally obtaining the approval of the editor of a journal that is certainly less highly rated, but which will finally agree to publish it! For a high IF journal, the time elapsed, from submission of the article to its acceptance, is on the order of four weeks for a study of high interest in the editor's "scope", which can reach six months, or even a year, for a work that is more distant from current events and the prevailing fashion or that requires additional validations. The team of researchers must therefore make a bet, while, at the same time, situating the journal's cursor in terms of IF: publish at a high level or make a date as quickly as possible by being satisfied with a lower IF…

The IFs of academic journals are ranked on a scale from 90 to 1. For example, the two most prestigious journals in science are *Nature* and *Science*, with an IF, in 2020, of 42.779 for *Nature* and 41.846 for *Science*; in medicine, they are *The Lancet* (IF $= 60.39$) and *The New England Journal of Medicine* (IF $= 91.245$). Works of broad scope are submitted to high IF journals; those that are of interest to a specialist readership, or that are syntheses of the literature, are generally published in journals with lower IF.

Acceptance of a publication in any journal follows the following steps: submission to the editor-in-chief, who, depending on the editorial policy of the journal, either rejects or accepts the paper for review. Two to three *reviewers* specialized in the subject of the submitted work are then appointed to evaluate the novelty of the proposed information and the accuracy of the scientific elements provided, as well as their coherence; these reviewers also verify the adequacy, in their opinion, between the content of the article and the impact factor of the journal.

In all, the evaluation takes between one week and three months, depending on the importance of the message and its urgency to be published.

Additional reviews may be requested, delaying final acceptance by the publisher and publication of the article.

Each researcher is both an author of peer-reviewed articles and a *reviewer* of other authors' articles. This makes the system appear somewhat schizophrenic in its current functioning, insofar as each researcher will
 (i) as an author, complain about severe *reviewers*, who criticize too harshly, if not reject, his/her manuscript;
(ii) criticize or even reject the articles of others at the same time, as a reviewer.
Thus, it often happens that, by being impossible to satisfy, intransigent, and arrogant with an author, a *reviewer* proceeds, from a psychological point of view, to a "negative transfer". In so doing, he or she is transferring onto others the suffering that he or she has endured as a researcher whose articles had previously been severely criticized or even rejected…!

Everyone will understand by this point: the current *peer review* system—the evaluation of researchers by their peers—has real flaws.

d. **The charismatic power of Key Opinion Leaders (KOLs)**

It is also necessary to take into account the charismatic power exercised by certain opinion leaders in any field.

The sociologist Max Weber (1864–1920) defines charismatic domination as "*the authority based on the personal and extraordinary grace of an individual*", of a leader. The community of followers trusts her/him, by emotional adhesion, and recognizes her/his capacity to proclaim legitimate sentences and, therefore, to "enact law".

In the field of medical research, this domination is reflected by the ability of certain scientists to promote theories or ideas that are likely to attract support:

 (i) of the majority of specialists on a subject, at a time when experiments cannot always definitively validate this or that hypothesis;
(ii) and, even more so, from the rest of the population, which, in this case, has no particular expertise and naturally trusts those who are supposed to have objective knowledge on this or that subject.

In this "grey zone" of scientific progress, one of the factors favoring the charismatic domination of certain scientists today is the increasingly specialized character of objective knowledge.

Thus, the general public, guided by all the actors of the popularization of scientific knowledge—journalists, politicians, philosophers, communication specialists, etc.—often recognizes a particular legitimacy in one or a few researchers who have an expertise in a particular scientific field.

This particular legitimacy distinguishes them from other researchers, making them *Key Opinion Leaders* who enjoy an important status within their own community. These *Key Opinion Leaders* can be found in a wide variety of scientific fields—biology, physics, mathematics, humanities and social sciences, etc. Their influence is often essential in the dissemination of scientific knowledge; their opinions are always listened to and very often followed.

However, this particular emphasis on *Key Opinion Leaders* sometimes leads to them being granted a specific expertise on a given subject that is, in fact, distinct from their own field of scientific competence. The result, in such circumstances, is a kind of "parasiting" of the scientific approach itself by various "preconceived notions": this is knowledge conceived before the scientific study, and often even directly drawn from empirical experience.

In the natural sciences, these "preconceived notions" can lead one to favor, without any logical basis, approaches that will ultimately prove to be unsuccessful, or to consider as vain or without a future experimental approaches that are nevertheless promising, and whose relevance and veracity will be discovered by the oecumene a few years later.

Until recently, the reception of mRNA technology by the majority of scientists—and *a fortiori* by the rest of the population—shows the characteristic features of this charismatic domination of opinion leaders, and the importance of these "preconceptions".

Thus, for many years, the dominant reaction of immunologists to the new mRNA technology and its relevance to health was doubt, skepticism, and even hostility, until early 2020. This also helps us to understand the relative conservatism of the review committees of the most prestigious scientific journals. This can result in the ignorance or misunderstanding of innovative work.

As a reminder, when CureVac started working on mRNA, Eli Gilboa's research on dendritic cells transfected with RNA was very popular among immunologists, despite the fact that the technology was very complex and involved very technical manipulations. Immunology journals and conferences regularly discussed it until the early 2010s; after that date, dendritic cell transfection was much less discussed (cf. *supra* Part II, Chap. 6, "Promising Studies from the 1990s").

In comparison, the injection of the mRNA itself into the organism, without prior cellular transfection, appeared to most specialists as a technology that was too simple, not to say simplistic, to work! For its detractors, there were three pejorative arguments:

- its cost—producing mRNA in vitro used to be expensive; this is no longer the case today;
- the weakness of the results, especially in the early 2000s. In experiments conducted between 2000 and 2002 at the Institut Pasteur, the results obtained using mRNA appeared less convincing than those with DNA. As a result, Institut Pasteur researchers did not pursue the mRNA route, in contrast to CureVac researchers, who were the only ones to grasp the full potential of this new vaccine technology;
- finally, many scientists and doctors thought that "mRNA cannot work, because it is a fragile molecule". This led them to disbelieve the first results obtained. Sometimes, prejudices have a hard skin and dominate the experiment. Not for investors!

Naturally, the situation changed completely with the appearance of the Covid-19 pandemic. Indeed, in January 2020, faced with the explosion of the Sars-CoV-2 infection around the world, and the urgency to fight this scourge, vaccination appeared to be the only way out. However, the first announcements of the imminence of this type of vaccine were perceived with skepticism: "*How could these vaccines already be available, when the specialists know very well that their marketing requires ten to fifteen years of work?*".

A number of scientists had heard of new mRNA vaccines, but few knew their status. Even among biologists, most were unaware of the biotechnology groundwork, its funding, and the nearly thirty years it took to bring mRNA vaccines to the forefront as the first prophylactic solution to the Covid pandemic!

And suddenly, more than twenty years of general "silence" were swept away by the announcement, made in March 2020 by Moderna and in April 2020 by Pfizer/BioNtech, of the first injections of anti-Covid mRNA vaccines. Amplified by the media, the anguish of political leaders in the face of the imminence of a second wave of Covid pandemic had finally imposed the reality of these vaccines on our planet!

To fully understand the success of these vaccines during the pandemic, it is necessary to first discuss the development of the three biotechs that have made them their *core business*—CureVac, BioNTech and Moderna.

References

1 Ugur Sahin, Katalin Kariko, and Özlem Türeci, "*MRNA-based therapeutics - developing a new class of drugs*", *Nature Reviews*, October 2014, Volume XIII, pp. 759–780. DOI: https://doi.org/10.1038/nrd4278.
2 Steve Pascolo, "*Messenger mRNA-based vaccines*", *Expert Opinion Biological Therapies,* August 4, 2004, pages 1285–1294. DOI: https://doi.org/10.1517/147125 984.8.1285.

17

The Development of Great mRNA Biotechs in the 2010s

Backed by unprecedented funding from private investors (venture capital), Curevac, BioNTech, and Moderna Therapeutics have written a veritable saga since 2000.

The context of fierce competition between these three companies, driven by high economic stakes, explains the advent of mRNA vaccines. From the 2010s onwards, each of them has experienced an impressive rise to prominence. Their history is well documented, and sources are numerous and widely available. They are presented in chronological order of their entry into the fray and their contribution to the field of mRNA vaccines.

CureVac

From its birth in 2000, in Tübingen, and during the following decade, CureVac focused its development strategy on mRNA vaccines (cf. *supra* Part II, Chap. 8, "The Birth of CureVac: The Era of the Pioneers").

The German biotech's growth continued in the 2010s, thanks to collaborations with various pharmaceutical companies (see https://web.archive.org/web/20150122153711/http://www.curevac.com/partnering/strategic-partnerships/, and also https://web.archive.org/web/20150122213016/http://www.curevac.com/research-development/ for clinical trials underway as of January 2015).

The following collaborations are particularly noteworthy:

© The Author(s), under exclusive license to Springer Nature   **133**
Switzerland AG 2023
J. Lemonnier and N. Lemonnier, *The Marathon of the Messenger*,
https://doi.org/10.1007/978-3-031-39300-6_17

- from October 2013, with Janssen Pharmaceuticals Inc, a subsidiary company of Johnson & Johnson, for the development of new flu vaccines;
- also in 2013, with the *Cancer Research Institute* and the Ludwig Institute for Cancer Research in New York, to initiate the clinical trial of new immunotherapy treatment options in lung cancer. CureVac would thus continue to work on the development of mRNA-based cancer vaccine strategies during these years;
- in March 2014, with Sanofi Pasteur and the French start-up In-Cell-Art; this collaboration was made possible by funding from DARPA. The *Defense Advanced Research Projects Agency* (DARPA) is an agency attached to the U.S. Department of Defense, with the objective of researching and developing new technologies for military use; it thus wished to fund a vaccine platform capable of reacting rapidly in the event of a microbiological terrorist attack. The goal was to develop a variety of vaccines using CureVac's mRNA technology, produced on the company's RNActive® platform. However, Sanofi Pasteur did not pursue this collaboration in the following years, choosing to focus on a different technology than mRNA;
- in March 2015, with the Bill-and-Melinda-Gates Foundation, for several projects to develop prophylactic vaccines with significant impact on population health in a number of countries. Initially, the project involved the development of mRNA vaccines against rotaviruses;
- finally, in September 2015, CureVac entered into a collaboration with the International AIDS Vaccine Initiative (IAVI) "*to accelerate the development of AIDS vaccines, utilizing novel immunogens developed by IAVI and partners, delivered* via *CureVac's novel messenger RNA (mRNA) technology*". As reported in the joint CureVac/IAVI press release, the Tübingen-based biotech appeared, in the mid-2010s, to be the company with the most experience in the use of therapeutic messenger RNA technology and the most ongoing clinical trials.

It should also be noted that, in March 2014, CureVac won a €2 million award from the European Commission for its innovative vaccine technology, which includes cold-chain-free transport and storage of vaccines in developing tropical countries. Because vaccines do not need to be refrigerated, the European Commission believes that CureVac's innovation can effectively contribute to improving public health in these countries.

CureVac today in no way resembles the biotech that it was at the beginning. Ingmar Hoerr and Florian Van der Mulbe are the last two remaining founders. Hoerr himself actually left CureVac in 2018; he would make a very brief return in March 2020 as the head of the company, before suffering a stroke that led CureVac to replace him only a few days after he took over. As of 2020, CureVac had approximately 500 employees.

BioNTech

BioNTech is a German biotechnology company founded in 2008 by two Turkish-born physicians, Ugur Sahin and his wife Özlem Türeci, and Christoph Huber, Professor of Oncology at the Johannes-Gutenberg University Mainz. BioNTech is a *spin-off* of the University of Mainz.

> In law, a *spin-off* is a commercial company created by the division of a larger company. In the case of biotechs, these are often commercial companies that are spun off from a university's training and research unit (UFR).

Ugur Sahin, a renowned oncologist, is steadfastly committed to bridging the gap between research and the clinic and has used mRNA, chronologically, first in oncology and then in treatment of infectious diseases.

Since the 2000s, he has been working on the development of personalized anti-cancer immunotherapies. In 2001, he and Özlem Türeci together created the company Ganymed Pharmaceuticals to develop monoclonal antibodies.

> Monoclonal antibodies are antibodies produced naturally by the same line of activated B cells or plasma cells, which recognize the same part of an antigen, or epitope. Whether or not the organism recognizes an epitope determines whether the antigen in question actually belongs to the "non-self" domain, which activates the production of antibodies (see *above*, Part I, Chap. 5, "The Immune Response").

In 2010, he initiated the project TRON—*Translationale Onkologie an des UniversitatMedizin Mainz* (Applied Oncology at Mainz Medical University). TRON is a platform for the development of cancer drugs based on immunotherapy, including mRNA cancer vaccines. In Ugur Sahin's mind, scientific advances in regard to cancer only make sense if they can be useful to patients, who expect effective and personalized responses from medicine. It turns out that the German term "übersetzend" (which translates into English as, appropriately enough, *"translational")* refers to the process of transferring basic scientific research into clinical practice. Indeed, there is sometimes a large gap between these two dimensions; the TRON project, initiated in 2010, had the essential goal of reducing this gap, or even making it disappear.

Since 2010, BioNTech has been working primarily on cancer mRNA vaccines. In oncology, the company is also developing DNA vaccines, therapeutic proteins, adjuvants, small molecule immunomodulators, and cell-based therapies primarily based on CAR T cells and antibodies (possibly encoded by non-immunogenic modified mRNA).

> *CAR-T cells* are T cells with a chimeric antigen *receptor*. This immunotherapy consists in taking the patient's T lymphocytes and then integrating a specific antigen that is referred to as "chimeric", because it is artificially created. The goal is to have these modified T cells recognize the patient's cancer cells, which will activate them to destroy them. Once modified, the T cells are reinjected into the patient.

From the outset, BioNTech has been a "horizontal" biotech: Ugur Sahin and Özlem Türeci have always wanted to ensure that all existing therapeutic solutions in oncology could be tested, optimized and experimented with in clinical trials.

All of BioNTech's senior managers are scientists, including the Director of Sales and the Chief Financial Officer. The BioNTech team consists of scientists from over 60 countries. Three renowned scientists should be mentioned: Sebastian Kreiter, Andreas Kuhn and Mustafa Diken, specializing in oncology, biochemistry and immunology, respectively.

Ugur Sahin and his collaborators were instrumental in the optimization of mRNA. Very soon after its founding, BioNTech became a leading mRNA vaccine company. It is therefore not surprising that their Covid vaccine was the first delivered to market, available for injection in the United States, on December 11, 2020.

Today, mRNA vaccinology represents the largest department at BioNTech, which clearly has the best expertise in the world in this field.

BioNTech is interested in both unmodified and modified mRNA: the company currently has a license to the University of Pennsylvania patent (see *supra* Part II, Chap. 10, "Modified and Unmodified mRNA: For What Purpose?"); in July 2013, it recruited one of the two inventors, Katalin Kariko, as Vice President of *Protein Replacement*. Modified mRNA is currently used for anti-Covid vaccination; it is also used in cancer immunotherapies: encoding bispecific antibodies, capable of binding to two different antigens, and which bind, for example, T or B lymphocytes to lymphoma or leukemia cancer cells.

BioNTech has been financially supported by the Strüngmann brothers, Thomas and Andreas, since its foundation in 2008. The details of the meeting between Ugur Sahin and Özlem Türeci on the one side and Thomas and Andreas Strüngmann on the other are described in detail in Joe Miller's book, *The Vaccine* [1]. The two billionaire brothers invested 150 million euros in the young biotech, an investment that ensured its financial stability for several years. Ugur Sahin and Özlem Türeci became 25% shareholders in BioNTech, and more importantly, they retained operational and strategic management. BioNTech can now develop as "*a stand-alone company, which will not be sold to a larger rival*" (page 203). As of 2009, BioNTech had 300 employees.

While BioNTech's early days were clearly easier than CureVac's, the Mainz-based biotech was having significant difficulty raising funds from investors in both Europe and the US in the early 2010s. This, in particular, led the Strüngmann brothers to reinvest in the company in 2011.

BioNTech finally managed to raise $270 million in late 2017. In July 2019, a second round of financing raised $325 million. What's more, in October 2019, BioNTech was listed for the first time on the U.S. technology exchange Nasdaq.

BioNTech also had to wait several years to enter into partnership relationships with pharmaceutical companies. After beginning a relationship with the American company Eli Lilly in 2015, Roche agreed to pay $310 million upfront to secure a 50/50 partnership with BioNTech [1]. And the American big pharma company Pfizer increasingly expressed interested in the German biotech, especially from 2017 on; thus, in July 2018, an agreement was signed with Pfizer on the development of an mRNA flu vaccine.

Finally, we note the signing, in September 2019, of an agreement with the Bill-and-Melinda-Gates Foundation to develop vaccines and immunotherapies against HIV and tuberculosis.

By early 2020, BioNTech had grown to 1300 employees.

Fig. 17.1 Stéphane Bancel

Moderna Therapeutics

The name Moderna, originally spelled ModeRNA, indicates, from the outset, the very spirit of the company: the exploitation of technologies based on modified RNA. Its president and CEO since 2011 is a Frenchman, Stéphane Bancel, an engineer who graduated from Supelec and was recruited when he was CEO of BioMérieux.

The company was founded in Boston before it had the technology to produce mRNA and conduct clinical trials. Unlike CureVac and BioNTech, Moderna received significant venture capital funding from its inception; its fundraising continued throughout the 2010s, indeed, it was able to obtain increasingly large sums, most notably, $1.4 billion between 2015 and 2018 (Fig. 17.1).

The company has also benefited from generous subsidies from the U.S. Government since its inception. In 2013, it received 25 million euros from DARPA [2].

For a long time, Moderna has been rather discreet about its biotechnological know-how and specifically targeted therapies.

During its first years of existence, between 2010 and 2014, Moderna worked solely on modified mRNA, which was thus presumed to be non-immunogenic, to express proteins without inducing an immune response. The information available on the company's website as of March 25, 2013, makes this clear: "*Moderna is creating first in class* in vivo *medicines known as messenger RNA Therapeutics*[TM], *which enable the body to produce its own healing proteins. This is a quantum change in the way protein therapeutics are produced and used, creating the possibility to treat unmet medical needs that cannot be addressed with current technologies. Our novel chemistry enables messenger RNA to elude the body's innate immune response. Once delivered, stable messenger RNA is translated into active, native protein by cells' natural, well-tuned machinery for protein production. messenger RNA Therapeutics*[TM] *enable the* in vivo *production of both intracellular proteins, which remain within the cells, and secreted proteins, which are released into the bloodstream and act to restore function elsewhere in the body*".[1]

In 2014–2015, Moderna differed from BioNTech in its therapeutic targets. It was applying the mRNA technology to various rare diseases by developing preclinical trials that proved disappointing.

From 2015 onwards, Moderna's strategic choices were focused on mRNA-based vaccination programs (anticancer or infectious diseases),[2] while it significantly increased its equity capital. The biotech then diversified its "mRNA" activities by joining forces with the pharmaceutical giant Astrazeneca to develop immuno-oncology programs based on mRNA. Moderna is financially responsible for research and preclinical development, while Astrazeneca is responsible for funding early human clinical trials.

In 2017, like BioNTech, the company ultimately chose to license the patent from the University of Pennsylvania for tens of millions of euros.

Moderna's success in the 2020s—at least that based on mRNA-modified therapies—has been achieved not in the field of protein replacement, but, of course, in the field of vaccination.

The details provided by Nathaniel Herzberg and Chloé Hecketsweiler in the French newspaper *Le Monde*, of November 30, 2020, are, in this respect, very interesting: "*Moderna (…) has an Achilles heel: it does not have the patent on the technology designed by UPenn to make a messenger RNA that is harmless to the body. Its scientists initially hoped to do as well, but all avenues explored*

---

[1] Cf. https://web.archive.org/web/20130411123321/http://www.modernatx.com/science.
[2] Cf. https://web.archive.org/web/20150829073316/http://www.modernatx.com/mrna-expression-platform.

ended in a dead end". "*We could see that UPenn's technology was better, but we didn't understand why*", admits Stéphane Bancel [3]. Moderna's "repositioning" clearly conceals the failures of its "replacement protein" strategy until 2017: indeed, until that date, Moderna's researchers had not succeeded in developing a synthetic mRNA that was not spotted by the patient's immune system. As developed earlier, this is indeed the reason for interest in pseudouridine-modified mRNA: used to be undetectable by the immune system, it does not block protein production (see Part II, Chap. 10 *supra*). The initial failure of Moderna's researchers thus paradoxically resulted in the incredible "success story" that we know from the Covid-19 pandemic. This success has opened up other therapeutic perspectives for Moderna in terms of replacement proteins (see Part II, Chap. 14 *above*, "Experiments and Clinical Trials in Other Therapeutic Fields").

As a result, Moderna's initial plans to focus on "modified mRNA (WITHOUT immune response)" therapies have given way to vaccine (WITH immune response) plans. In January 2022, Moderna announced that it had begun Phase 1 clinical trials of an mRNA-based AIDS vaccine.

Moderna has also had access to more facilities than BioNTech for the development of its mRNA vaccine for Covid. Stéphane Bancel's biotech partnered with the *National Institute of Allergy and Infectious Diseases*, a U.S. Government agency headed by Anthony Fauci [1]. Moreover, unlike BioNTech, Moderna was not required by U.S. regulators to conduct a toxicology study, since it had already tested this same formulation for another vaccine in 2019.

Moderna had 800 employees in 2019; as of July 2022, it had 3200.

Currently, these three biotechs produce their own mRNA to make their own drugs, including vaccines against Covid. They do not sell it. Only the Belgian biotech company Etherna currently sells its own unmodified mRNA, thus being a *service provider*. This *spin-off* from the Free University of Brussels sells unmodified mRNAs manufactured according to GMP production rules, for immunotherapy treatments of cancers and infectious diseases.

The CureVac, BioNTech and Moderna itineraries, while varied, share some common characteristics.

• The same economic development model

The development model of the three biotechs is identical: first of all, the use of private investors who could inject considerable sums into each company with a high level of risk (venture capital). Especially in the early years, the development of new molecules, as well as preclinical trials, requires significant investments that the researchers or professors working for these biotechs generally do not have.

• A second phase of collaboration with the pharmaceutical industry

When the start-up becomes known, after the first encouraging research results, collaborations with large pharmaceutical companies such as *AstraZeneca, GSK, Sanofi, Bayer, Pfizer*, etc., take place. Generally, *down payments* are agreed upon: the biotech finances the research work, the technological equipment and the preclinical trials. These "big pharma" companies pay for the clinical trials, particularly phase III, which can be extremely expensive when carried out on several tens of thousands of patients.

During these clinical trials, the investor or the large pharmaceutical company covers the cost per patient recruited (around 30,000 euros), the payment of medical expenses, the costs of audits and ethics committees responsible for ensuring that the clinical trials are carried out in accordance with the established protocols, and the possible costs of certain infrastructures. The total cost of a phase-III clinical trial can easily reach several tens of millions of euros and, in general, often amounts to several hundred million euros. However, the size of the sample depends on the frequency of the disease: for a rare disease, the phase III clinical trial will only involve a small number of patients.

These phase-III clinical trials also give rise to compensation from the directors of the participating hospitals or clinics: payment of sums of money and mentions in scientific and medical articles or at conferences.

Following the cooperation with these "big pharma" companies, the biotech is usually listed on the stock exchange. If the phase III clinical trial is successful, the biotech receives a royalty on drug sales, for example, 5%. It is at this point that investors recover, to a very significant degree, their initial investment.

• The need for industrial-scale mRNA production

The biotech must propose industrialist production processes that are in compliance with GMP standards. The initial programs of CureVac, Moderna and BioNTech only planned to use a few grams of mRNA per year, with the objective of manufacturing small quantities of drugs for the treatment of certain rare diseases.

COVID-19 changed everything in these projects and objectives: indeed, it was necessary to set up very quickly a significant production of mRNA, on a kilogram scale, and always, naturally, in compliance with GMP standards. However, since such industrial-scale mRNA production was rapid and did not pose any real problems, CureVac, BioNTech and Moderna felt that the development of an mRNA vaccine would be less risky and less expensive than the development of a recombinant protein vaccine.

At present, these three biotechs each have their own processes for developing and producing mRNA on a large scale. These are protected by a number of patents that provide them with licensing revenues (see *below*, Part III, Chap. 18, "The Importance of Intellectual Property and Patents").

Note that, in the case of such large-scale mRNA production, the purification method can be a little less advanced than in the case of a small-scale production: no recourse to the classical HPLC purification process to obtain an mRNA that is certainly very pure, but which is only applicable for the production of small quantities. Indeed, if one chose an identical purification, it would be necessary to wait months before obtaining the desired industrial quantity.

Under the conditions of industrial production of mRNA in small quantities, one can proceed to the elimination of all the aberrant mRNAs, those that are too long or too short. But for the industrial production of mRNA in large quantities, one proceeds only to the elimination of all the molecules that are not mRNA. On the *other hand,* contaminating mRNAs that are smaller or larger than the desired one are kept. It is therefore very likely that more "truncated" mRNAs will remain that are not suitable for translating the desired protein.

However, this does not affect the quality of the resulting vaccine. And, in fact, in the development of the Covid vaccines, the right ratio of good quality mRNA for the intended purpose has not been a major issue, especially for the agencies responsible for licensing and marketing these new mRNA vaccines. As Steve Pascolo reports, "*In BioNTech's early trials, they had immune responses with just one microgram of RNA (…) Now the vaccine is supposed to have 30 µg, so they have room to spare*". [4].

Finally, each of the three biotechs has specific patents that, being at the heart of their economic development model, constitute an important part of their revenues and play an essential role in their economic sustainability.

The following will demonstrate this.

References

1 Joe Miller, with Özlem Türeci and Ugur Sahin, "*The Vaccine. Inside the Race to Conquer the Covid-19 Pandemic*", February 2022, St. Martin's Press.
2 "DARPA awards Moderna Therapeutics a grant for up to $25 million to develop Messenger RNA Therapeutics", Moderna, October 2, 2013, https://www.fierce biotech.com/biotech/darpa-awards-moderna-therapeutics-a-grant-for-up-to-25-million-to-develop-messenger-rna.
3 Nathaniel Herzberg and Chloé Hecketsweiler, "Covid-19: the saga of the messenger RNA vaccine now in the final sprint," *Le Monde,* 30 November 2020.
4 Steve Pascolo, in Lise Barnéoud, "*Ce que disent les documents sur les vaccins anti-Covid-19 volés à l'Agence européenne des médicaments*", in *Le Monde*, 16 January 2021. Article available at: https://www.lemonde.fr/planete/article/2021/01/16/vaccins-ce-que-disent-les-documents-voles-a-l-agence-europeenne-des-medicaments_6066502_3244.html.

18

The Importance of Intellectual Property and Patents

Revenues from patents, which cover the inventions and innovations of biotechnology companies, are essential to their sustainability. This is an obvious reason to understand the importance of intellectual property in the development of mRNA therapeutics by CureVac, BioNTech and Moderna.

In France, intellectual property protection for inventions and industrial innovations is provided by Article L. 611-10 of the Code de la Propriété Intellectuelle (Intellectual Property Code), which states (paragraph 1): "*New inventions involving an inventive step and capable of industrial application are patentable in all fields of technology*". As specified in Article L. 611-2 of the same code, this protection lasts for twenty years for patents, starting from the date of filing of the application.

The purpose of a patent is to protect a technical innovation, i.e., a product or process that provides a new solution to a given technical problem. This assumes that the product or process is new, or significantly improved compared to what is already available on the market.

> The 1992 *Oslo Manual of the OECD*, which brings together "*guidelines for the collection and interpretation of data on innovation relating to the definition of the scope of research and development (R&D) activities*", distinguishes four categories of innovation: product innovations, process innovations, marketing innovations and organizational innovations. The patents of biotechs are very generally aimed at product or process innovations, defined as follows by the Manual: "*A technological product innovation is defined as the development/*

© The Author(s), under exclusive license to Springer Nature Switzerland AG 2023
J. Lemonnier and N. Lemonnier, *The Marathon of the Messenger*, https://doi.org/10.1007/978-3-031-39300-6_18

marketing of a more efficient product with the aim of providing the consumer with objectively new or improved services. Technological process innovation means the development/adoption of new or significantly improved production or distribution methods. It may involve changes affecting—separately or simultaneously—materials, human resources or work methods". For industrial companies such as biotechs, these two types of innovation perfectly fit into their investment strategies and decisively contribute to their industrial and commercial development. The comparison with the theory of the economist Joseph Schumpeter, who insists on innovation as a "commercial experiment" provoking an in-depth restructuring of industries and markets, appears, in this case, to be particularly relevant for biotechnology companies.

Nowadays, patents always belong to the companies, universities and other organizations for which the inventors work. However, the inventor(s) of the patent are always compensated:

(i) Either before the invention itself: the researcher receives a bonus before the patent is filed, but does not receive any income from royalties afterwards, i.e., income from patent licenses;

(ii) or after: the researcher receives additional remuneration provided for in his or her employment contract, as well as income linked to his or her personal profit-sharing in the royalties received by the company. This is a point to which particular importance should be attached. In France, depending on the research organization to which they are attached, researchers may be more or less remunerated by the filing of a patent that they have helped to write. The vision of the necessarily disinterested researcher can be counterproductive insofar as it is not an incentive. The percentage of remuneration of the researcher, whether he/she contributes personally or as a means of financing his/her department (laboratory for biology), is an important parameter, but in France, it is unfortunately too often neglected. In contrast, an American laboratory manager, once private or public funding has been obtained, can decide how to allocate his/her budget.

In fact, the regimes vary from country to country; the incentives for inventors also vary greatly, depending on the intellectual property protection strategies of the patent-owning organizations.

Let us now look at the case of France to determine:

(i) how patents are granted to companies, universities or other organizations that own them;

(ii) and how they concretely make possible the protection of inventions, in particular, in the field of biotechnology.

In France, patent protection is provided by an independent organization, the INPI-Institut National de la Propriété Industrielle (National Institute of Industrial Property). The INPI issues industrial property titles, such as patents.

Pharmaceutical patents are special patents, as stated on the INPI website: "Pharmaceutical patents are granted, like all other patents, for a period of twenty years from the date of filing and in return for the payment of annual fees. However, pharmaceutical products require a marketing authorization (MA) in order to be marketed. This authorization is generally not issued for several years. Therefore, to compensate for this period during which the patent cannot be exploited, a special title has been created, the Supplementary Protection Certificate (SPC), which extends the rights of the owner of a patent on a pharmaceutical product for up to five additional years".[1]

Other organizations grant patents at an international level: for Europe, the European Patent Office (EPO), and for the world, the World Intellectual Property Organization (WIPO). With the latter organization, it is possible to seek patent protection simultaneously in many countries by filing a single international application.

The development of new patents and the revenues from licensing them to other companies represent an essential strategic development axis for all biotechs in general, and for those that have based their business model on mRNA vaccination in particular.

When a biotech company files a patent, it is very common for it to be attacked by other biotechs, who may rely on a relatively imprecise term formulated five or ten years earlier to challenge the novelty of the patent. This leads to lengthy and often costly litigation.

Patent prosecution usually takes several years, which means that the filing of a patent is not the same as validation. Patent analysts at WIPO or the EPO usually start by attacking the patent as not being new. Since patents are sometimes on very specific subjects, analysts may be tempted to do keyword searches, using computer algorithms. This method of analysis thus quite often helps, for a given patent, in finding the same word in a previous patent, which, however, relates to something completely different, even a very different field of research.

---

[1] More information on intellectual property is available at https://www.inpi.fr/fr/comprendre-la-pro priete-intellectuelle/le-brevet.

It is therefore sometimes difficult for a biotech company to have the innovative character of a patent recognized, when it finds itself corresponding with a person whose scientific field is sometimes very distant: it can thus be very difficult for a reviewer (evaluator) who has, for example, a physicist's training to fully understand a patent filed by a biotech company, and for the biotech to convince them that it is indeed an innovative product or process.

Patents include "*claims*". This is the essential part of the patent. The first *claims* are very general and are not often recognized as innovative. The last *claims* of the patent are the most important: they represent the most specific part of the innovation.

The time required to process a patent varies greatly, depending on the nature of the patent filed. Let's take a real-life example: a patent is filed for the development of a six-wheeled car in the color red; it will be validated very quickly, as the verification of the possible anteriority of another manufacturer will take very little time. In this case, the scope of the patent will be extremely limited: nothing will prevent another company from filing a patent for the same product, but in orange.

Naturally, for patents filed by biotechs, the delays are much longer. Thanks to a specific coding, one can follow the status of the patent on the WIPO or EPO websites: "filed", "validated", "attacked", etc.

During the interval between the filing of a patent and its validation, there is nothing to prevent another company from developing a product that falls under a particular *claim*. However, if the patent is validated, that company will have to pay a license fee.

The choice to industrialize this or that product is not always based on science; quite often, legal considerations, linked to the protection of industrial property, prevail. In fact, this responds to the general objective of companies to make a profit.

It should also be noted that physicians are generally unaware of the more effective products developed by biotechs; in fact, they use what the pharmaceutical industry has developed. They are therefore not aware of new products developed by biotech companies, which are in clinical trials (even in phase III), whether or not they are patentable products.

For the specific case of mRNA, consider a biotech company that has two types of mRNA:

(i) an unpatented mRNA A that works very well;
(ii) a patented B mRNA that works a little less well.

In the short term, it will develop mRNA B. In fact, at first, it will not earn anything from making industrial investments, and will eventually make its unpatented mRNA A available to all its competitors. The products marketed by this or that biotech will therefore not necessarily be the best of the moment, at least initially.

However, the reasoning in the medium and long term will be different: for any improvement made to a patentable product, the biotech will have, over a longer time scale, an incentive to work more on mRNA A than on mRNA B. Indeed, mRNA A is the product that will give it a truly advantageous position in a competitive environment. The challenge for the company is to develop a patentable innovation that will make its mRNA A appear to be a "technological breakthrough" product.

Let us consider this problem in light of the previous development in regard to modified mRNA and unmodified mRNA (see *supra* Part II, Chap. 10, "Modified mRNA and Unmodified mRNA: To What End?").

At present, it is possible that modified, patent-protected mRNA is no better than unmodified, unprotected mRNA. Eventually, however, it is certain that, in a competitive biotech environment, any innovation that gives one or the other a decisive advantage in mRNA optimization (modified or unmodified) will eventually win out decisively over competing innovations.

Clinical trials being the judge of peace, the decisive criterion is therefore that of the therapeutic virtues of this or that mRNA to effectively treat this or that pathology. In the corresponding specific market, the logical consequence is that there should eventually be only one type of mRNA, the one with proven superior therapeutic efficacy. In economic terms, this is therefore a "medical asset", whose fundamental value over time is the subject of prospective anticipations.

Today, the mRNAs developed by BioNTech, Moderna and CureVac are subject to intellectual property protection. They cover the following elements:

- 3' UTR and 5' UTR (different for BioNTech, Moderna and CureVac);
- mRNA production steps (especially related to purification);
- formulation differences (using Acuitas Therapeutics' products, if applicable) (see *above*, Part II, Chap. 9, "Solutions for mRNA Optimization");
- the use of pseudo-uridine for BioNTech and Moderna;
- freeze-drying processes (owned by, among other companies, CureVac);
- processes for using mRNA (e.g., to encode antibodies or antigens).

In particular, CureVac has filed numerous "messenger RNA-encoded anti-body" patents since 2008. This multitude of patents is not surprising, as the human antibody repertoire includes 100,000,000,000 different specific anti-bodies (targeting as many antigens), even though there are five main classes of antibodies, which are very similar in construction.

In CureVac's first patent from 2008 [1], the RNA involved can of course be mRNA (see *claim* 1). This patent was approved in 2015, and then immediately attacked, hence its "pending" status. It was indeed an invention with huge financial consequences, as it protected, in particular, the replacement of all monoclonal antibodies by mRNA.

As evidenced, in this case, by a July 2017 *Nature* article by Ugur Sahin's team [2], BioNTech was also developing messenger RNA-encoded bispecific antibodies at the same time, with an anti-cancer perspective. However, we do not have confirmation that BioNTech was indeed among the companies that attacked the CureVac patent approved in 2015.

BioNTech, on the other hand, has developed a number of "RNA-encoded tumor antigen" patents, particularly since 2012 [3].

In addition, the use of Trilink's CleanCap for 5′ cap optimization (see *above*, Part II, Chap. 9, "Solutions for mRNA Optimization") requires the three biotechs to pay licenses.

Regarding the formulation of mRNA, the patent filed in 1992 by Frédéric Martinon and Pierre Meulien [4] using liposomes forced CureVac, in 2000, to propose the use of protamine. It was not until 2012 that CureVac turned to lipid nanoparticles. For its part, BioNTech perfected the liposomes used as early as 2010, and filed several patents, thanks to the work of its chemist, Heinrich Haas, now Vice President of "RNA Formulation and Drug Delivery."

To summarize, each biotech currently has several "layers" of patents, with which it can protect its innovations and continue its development.

Thus, considering this protection by patents, one can deplore the excessive length and importance of legal procedures, as well as the inappropriate industrialization of products that, in the short term, will not be the best available. However, for all biotechs, intellectual property is now a key success factor. By differentiating competitors in the medium and long term, it is a source of competition and, consequently, of permanent therapeutic innovations.

Patent disputes between competitors are common, often resulting in legal action, particularly in the United States. This is, of course, also true for biotechs that have worked on mRNA.

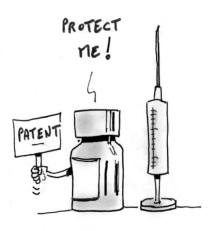

A very significant example is the lawsuit filed on July 5, 2022, by CureVac against BioNTech for intellectual property infringement. Remarkably, this is a dispute between two German companies, and it is especially remarkable that it is between the two biotechs that have played the most significant role in the development of mRNA vaccines. As journalist Monika Dunkel reported in the July 6, 2022 issue of *Capital* [5]: "*CureVac cites four of its own patents that Biontech allegedly infringes. These are processes for the production of mRNA molecules that increased the stability of messenger RNA, as well as the formulation of mRNA vaccines*" ("*Konkret nennt Curevac vier eigene Patente gegen die Biontech verstoßen haben soll. Es geht um Verfahren zur Herstellung von mRNA-Molekülen, die die Stabilität der Boten-RNA erhöht haben, außerdem um die mRNA-Impfstoffformulierung*"). Is this a "sore loser" attitude on the part of CureVac, which has not been able to develop and market an effective anti-Covid mRNA vaccine (cf. *infra* Part III, Chap. 19, "2020: The Triumph of Anti-Covid mRNA Vaccines")? Is it a more global problem of lack of scientific innovation or inspiration at CureVac, which this litigation is supposed to mask or, on the contrary, reveal? Or, contrastingly, a desire to remind us that CureVac was indeed the first biotech company founded on the use and development of mRNA technology? Lawyers, and later historians, will decide. In any case, this confirms the fundamental importance of patents for the development of the most innovative biotechs.

On August 26, 2022, Moderna also decided to sue BioNTech and Pfizer, accusing them, in a press release [6], of infringing upon its intellectual property in regard to messenger RNA vaccines. In particular, the infringement allegedly involves two patents:

I. "*The chemical modification of mRNA*," which, in the Comirnaty vaccine, "*is exactly the same as in Spikevax®*", refers here to the pseudo-uridine modification of mRNA discussed above (see Part II, Chap. 10, "Modified Versus Unmodified mRNA: For What Purpose?" and Part II, Chap. 15, "Modified Versus Unmodified mRNA: Not just a Scientific Issue"). As noted, this mRNA modification process has been sublicensed to both BioNTech and Moderna.

II. Moreover, according to Moderna's press release, "*Pfizer and BioNTech copied Moderna's approach to encoding the entire spike protein in a lipid nanoparticle formulation for a coronavirus*," an approach Moderna scientists had also developed in another vaccine several years before Covid 19 emerged. This specific point actually refers to BioNTech's choice of the BNT162b2 vaccine in the fight against Covid, a vaccine based on the coding of the entire Spike protein. Details of BioNTech's various clinical trials, which led to the choice of the BNT162b2 vaccine, are provided *below* in Part III, Chap. 19, "The Triumph of Covid mRNA Vaccines".

The current patent system is regulated by their limited duration: no biotech can live on its achievements. This is a tremendous incentive for curiosity, creativity and innovation. Thus, each company grows, diversifies and prospers, or even disappears: the licenses from its patents guarantee revenues directly proportional to the innovation. In a competitive environment, success goes to the best products, for the same reasons that natural selection favors the organisms best adapted to their environment.

For this reason, the proposal made on May 5, 2021, by the US President, Joe Biden, to lift the patents on the mRNA vaccines developed by BioNTech, Moderna and CureVac appears to be nothing more than a political communication stunt. In reality, it is highly unlikely, and highly undesirable, that any government would actually decide to implement such a measure. Its immediate consequence would be to discourage private investment in biotech companies, making the prospects of a return on investment unlikely. In the longer term, it would penalize any progress and innovation in the development of mRNA therapies, and ultimately prevent users from accessing the most effective vaccine at the most reasonable cost.

Generally speaking, the States only have a regulatory role, through the drug agencies (particularly the EMA and the FDA).

The benefits of free competition between firms in different markets are well expressed by the economist Jean Tirole: "*Competition does not just mean lower prices. It pushes firms to produce more efficiently and to innovate. It promotes diversity of approaches and experiences, leading to the emergence of more efficient technological choices and economic models, as we see today in the Internet*" ("*La concurrence ne se traduit pas seulement par des prix plus bas. Elle pousse les entreprises à produire de façon plus efficace et à innover. Elle promeut la diversité des approches et des expériences, faisant émerger des choix technologiques et des modèles économiques plus performants, comme on le voit aujourd'hui dans l'Internet*") [7].

In this sense, the rapid development of vaccines against Covid also appears to be the positive effect of this free competition.

References

1 Ingmarr Hoerr, Jochen Probst, Steve Pascolo, "*Rna-coded antibody*, German patent (CureVac)", No. US20100189729A1, filing date: 8 January 2008, publication date: 21 December 2009. Available at: https://patents.google.com/patent/US20100189729A1/en?oq=(US2010189729A1+RNA-CODED+ANTIBODY.

2 Christiane R Stadler, Hayat Bähr-Mahmud, Leyla Celik, Bernhard Hebich, Alexandra S Roth, René P Roth, Katalin Karikó, Özlem Türeci, Ugur Sahin,

"Elimination of large tumors in mice by mRNA-encoded bispecific antibodies", *Nature Medicine*, 2017 Jul;23(7):815–817. Doi: https://doi.org/10.1038/nm. 4356. EPUB 2017 Jun 12.

3 Ugur Sahin, Heinrich Haas, Sebastian Kreiter, Mustafa DIKEN, Daniel Fritz, Martin MENG, Lena Mareen KRANZ, Kerstin REUTER, *"Rna formulation for immunotherapy"*, German patent (BioNTech), No. EP2830593B1, filing date: March 26, 2012, publication date: October 3, 2013. Available at: https://pat ents.google.com/patent/WO2013143555A1/en.

4 Pierre Meulien, Shivadasan Krishnan, Frédéric Martinon, «*RNA Delivery Vector*», N°. WO1992019752, filling date: 30 April 1992, publication Date: 12 November 1992. Available at: https://patentscope.wipo.int/search/fr/detail.jsf?docId=WO1 992019752&_cid=P20-LBGRFK-99288-1.

5 Monika Dunkel, *"CureVac vs. BioNTech: a useful argument,"* *"Curevac kontra Biontech: Ein nützlicher Streit"*, in *Capital*, July 6, 2022. Article available at https://www.capital.de/amp/wirtschaft-politik/curevac-kontra-biontech--ein-nuetzlicher-streit-32514850.html.

6 *"Moderna Sues Pfizer and BioNTech for Infringing Patents Central to Moderna's Innovative mRNA Technology Platform"*, Moderna Press Release, August 26, 2022, available at https://investors.modernatx.com/news/news-details/2022/ Moderna-Sues-Pfizer-and-BioNTech-for-Infringing-Patents-Central-to-Mod ernas-Innovative-mRNA-Technology-Platform/default.aspx.

7 Jean Tirole, *Économie du bien commun*, Éd. PUF, 2016, p. 473.

19

2020: The Triumph of Anti-Covid mRNA Vaccines

Starting in the fall of 2019, the story of mRNA vaccines becomes closely tied to the emergence of the coronavirus pandemic. This is the "tip of the iceberg", amplified today by all media. So, let us now take a look at the main steps of this development of these anti-Covid mRNA vaccines, and their main characteristics.

First identified in November 2019 in central China's Hubei province of Wuhan, the virus then spread rapidly around the world during the winter. The World Health Organization (WHO) declared a state of international public health emergency on January 30, 2020, before declaring Covid-19 a pandemic on March 11, 2020, with much delay.

The virus responsible is a coronavirus (Sars-Cov-2), of a viral family with a spicules-crowned envelope, whose genetic sequence was published in *Nature* on January 11, 2020 [Fig. 19.1]. It appears to be similar to the Sars-CoV that appeared in 2002 - 2003, with which it shares 80% of its genetic heritage.

> The **severe acute** *respiratory* **syndrome coronavirus** or **Sars-CoV** (sometimes **Sars-CoV-1,** to differentiate it from Sars-CoV-2, which appeared in 2019) is the coronavirus responsible for the severe acute respiratory syndrome (SARS) epidemic that occurred from 2002 to 2004. This infectious agent appeared in November 2002, in Guangdong province, China. Between November 1, 2002, and August 31, 2003, the virus infected 8,096 people in some 30 countries, causing 774 deaths, mainly in China, Hong Kong, Taiwan, and Southeast Asia.

© The Author(s), under exclusive license to Springer Nature Switzerland AG 2023
J. Lemonnier and N. Lemonnier, *The Marathon of the Messenger*,
https://doi.org/10.1007/978-3-031-39300-6_19

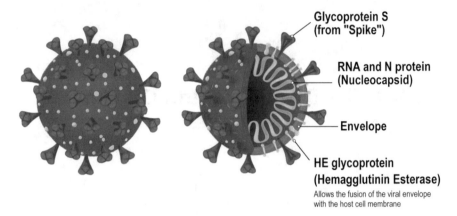

Glycoprotein S
(from "Spike")

RNA and N protein
(Nucleocapsid)

Envelope

HE glycoprotein
(Hemagglutinin Esterase)
Allows the fusion of the viral envelope
with the host cell membrane

Fig. 19.1 SARS-CoV-2—A Simplified Scheme

This sequence homology has focused the interest of infectious disease specialists on a spike envelope protein, the S protein (for "spike").

Previous studies on Sars-CoV had revealed the mechanism of viral infection: the S protein recognizes the ACE2 membrane receptor of epithelial cells, particularly nasopharyngeal cells. These studies also demonstrated that the specific conformation of the S protein made it susceptible to being recognized by antibodies that neutralized the virus.

ROLE OF THE
SPIKE PROTEIN.

Spike is a large transmembrane protein containing 1273 amino acids. In detail, the conformation in question is called a prefusion. BioNTech, Moderna and CureVac used their pre-existing knowledge to design their coding sequence. To stabilize this conformation in the manufacture of the spike mRNA vaccine, two consecutive prolines in the C-terminal portion of the protein replace the natural amino acids (KV sequence at positions 986 and 987 of the protein). The use of these two proline amino acids at the carboxyl end of the spike protein thus promotes the "prefusion" structure of the Sars-CoV-2 spike expressed by the in vitro transcribed mRNA on the surface of transfected cells.

BioNTech, CureVac and Moderna used the mRNA encoding the S protein as a tool for an anti-Covid-19 vaccine strategy! By early 2020, the three biotechs were ready to provide the expertise and infrastructure needed to produce a prophylactic mRNA vaccine.

Thus, Moderna's and BioNTech's initial work began in January 2020; their mRNA vaccines were designed in the spring and marketed by the end of the year. While phase-III clinical trials continued until October 2022 for Moderna, and until May 2023 for Pfizer/BioNTech, the testing and the data collection phase that concluded the efficacy of these two vaccines were completed in December 2020. For any type of study, it is necessary to collect as much data as possible in order to identify the vaccine as precisely as possible.

Figure 19.2 describes the mode of action of the anti-Covid mRNA vaccine on our bodies. In particular, it reports on the induced immune response.

The comparison table below (Table 19.1) shows the common features and differences of the three anti-Covid mRNA vaccines.

Finally, for more information on these vaccines, it is useful to refer to the box "Additional Data on Covid mRNA Vaccines" below.

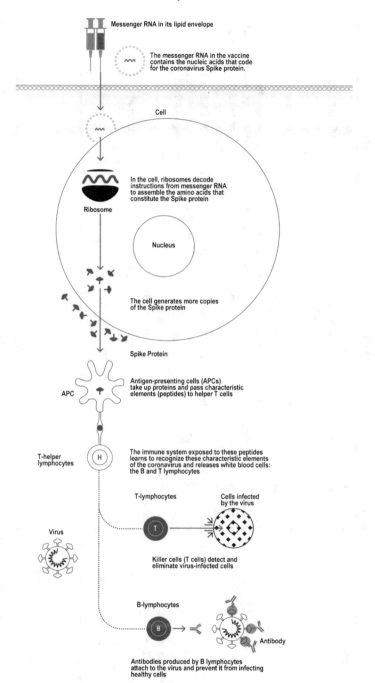

Fig. 19.2 The action of mRNA vaccine on your body

Table 19.1 The characteristics of three mRNA vaccines

	BioNTech/Pfizer	Moderna	CureVac
Name of the vaccine	Comirnaty	Spikevax	CVnCoV
Type of mRNA	mRNA modified by pseudo-uridine	mRNA modified by pseudo-uridine	Unmodified mRNA
Coded viral protein	Full spike	Full spike	Full spike
Concentration	30 μg of mRNA in 0.3 mg of saline solution	100 μg of mRNA in 0.5 mg of saline solution	12 microgrammes d'ARNm
Composition	Vegan—very few additives (water, salts, sucrose and lipid nanoparticles)—no aluminium	Vegan—very few additives (water, salts, sucrose and lipid nanoparticles)—no aluminium	Vegan—very few additives (water, salts, sucrose and lipid nanoparticles)—no aluminium
Clinical trials	April 23, 2020 (start of phase I)—May 2, 2023 (end of phase III)	March 16, 2020 (start of phase I)—October 27, 2022 (end of phase III)	June 19, 2020 (start of phase I)—June 2021 (end of phase III)
FDA approval date[1]	December 11, 2020	December 18, 2020	
EMA approval date[2]	December 21, 2020	January 6; 2021	
Injection mode	Intramuscular	Intramuscular	Intramuscular
Number of injections	2	2	?
Protection rate after both doses (%)	95	95	48

[1] The *Food and Drug Administration* (FDA) is the U.S. Government agency that authorizes the marketing and sale of drugs and vaccines in the United States.

[2] The *European Medicines Agency* (EMA) is the European administration that authorizes the marketing and sale of medicines and vaccines in the European Union. Its authority is exercised through national agencies, in particular, for France, the Agence nationale de sécurité du médicament et des produits de santé (ANSM).

	BioNTech/Pfizer	Moderna	CureVac
Time between the two injections	Theoretically, three or four weeks	Theoretically, four weeks	
Contraindications	No	No	
Possible side effects	Mild, in almost all cases (fatigue, chills, headache, myalgia and pain at the injection site)	Mild, but more severe than BioNTech/Pfizer, in almost all cases (fatigue, chills, headache, myalgia and pain at the injection site)	
Duration of vaccine immunity	At least six months	At least six months	
Normal storage temperature	− 70 °C	− 20 °C	
Shelf life in refrigerator (5 °C)	5 days	30 days	
Price of the two doses in Europe	19.50 € (mid-August 2021)	21.50 € (mid-August 2021)	
Reimbursement by the French Social Security	Yes	Yes	

19.1   Additional Data on Anti-Covid mRNA Vaccines

The anti-Covid mRNA vaccines have some specific characteristics, which are worth highlighting.

Depending on the country, the time between the two injections can vary

At the time that they were marketing their vaccines, Pfizer/BioNTech and Moderna were insistent about the time between the two injections. Partial protection is already seen 14 days after the first injection; the very significant decrease in the number of new cases seen in Israel and Great Britain, as early as February 2021, showed that significant vaccine immunity is already provided by the first injection.

Near total protection is achieved one week after the second injection. Given these characteristics, public health policies in different countries have provided different answers to the question: is it better to partially protect more individuals or totally protect fewer individuals? The United Kingdom, for example, has favored the former approach, while France has favored the latter.

The absence of contraindications

Three medical contraindications to the two mRNA vaccines already on the market have been formulated in France by the Ministry of Health:

- a history of pediatric inflammatory multisystemic syndrome (PIMS), an extremely rare complication that has affected some children and adolescents following Sars-CoV-2 infection;
- a history of severe myocarditis, pericarditis, or hepatitis that required hospitalization and follows a first injection of mRNA vaccine. Exceptional cases of myocarditis and pericarditis have been reported following injection of an mRNA vaccine (Pfizer/BioNTech or Moderna). In all cases concerned, the evolution was favorable;
- an allergy to one of the components of the vaccine, PEG 2000 (or polyethylene glycol), used to increase both the stability of liposomes and their circulation time in the body; this type of allergy would come to concern about 10 cases in France, according to the Ministry of Health.

However, no diseases, not even autoimmune disease, diabetes, cancer, or acquired immunodeficiency syndrome, were found to be contraindications to vaccination during the clinical trials. Similarly, no increased risk was found in pregnant or lactating women. As for allergic hypersensitivity in some people, it was not systematically contraindicated to vaccination, because of the reassuring observations made during the clinical trials. In France, however, and in practice, the administration of an mRNA vaccine must take place in a facility capable of managing a severe immediate hypersensitivity reaction.

Duration of vaccine immunity

The data, available in retrospect, show that vaccine immunity extends beyond six or seven months. Vaccine boosters may be necessary, given the persistence of neutralizing antibodies and the appearance of antibodies after new exposure to the virus (implying mobilization of memory B cells) (see Part I, Chap. 5 *above*, "The Immune Response"). The emergence of variants, a factor of escape from vaccination, may also justify booster vaccinations. Under these conditions, monitoring of vaccine and natural immunity is necessary.

Additional details on BioNTech's clinical trials

BioNTech, with a wide range of in vitro transcribed mRNA technologies, has developed four different mRNA vaccines as of January 2020 (see *supra* Part II, Chap. 15, "Modified mRNA and Unmodified mRNA: Not just a Scientific Challenge"):

- an mRNA coding for spike in a full-length prefusion conformation with pseudo-uridines (BNT162b2);
- two mRNAs encoding only the "receptor binding" domain of the spike protein (since antibodies targeting this domain alone may be more effective and also reduce the theoretical possibility of induction of infection-facilitating antibodies), one with pseudo-uridines (BNT162b1) and the other with unmodified uridines (BNT162a1);
- a self-amplifying mRNA (since this format requires much lower doses than the non-amplifying mRNA) encoding the whole spike (BNT162c2).

After initiating the first injection on April 23, 2020, BioNTech finally decided to use the pseudouridine-modified mRNA encoding spike in the "prefusion" conformation (BNT162b2) in the Phase III trial.

Raw materials required for manufacturing in industrial quantities

All three major players already had production facilities to manufacture in vitro transcribed mRNA under GMP pharmaceutical conditions.

However, mass production of a prophylactic vaccine in a short time frame is a challenge. If one billion doses are ultimately needed urgently (to vaccinate 500 million people twice, covering about 10% of the population: for example, people at risk and health care workers), then 100 kg of mRNA is needed for Moderna, 30 kg of mRNA for BioNTech and 12 kg of mRNA for CureVac. According to data reported by Steve Pascolo [1], the in vitro transcript volumes are then at least 20,000, 6000 and 2400 L, respectively (assuming a concentration of at least 5 mg/ml mRNA at the end of the transcription reaction).

The availability of raw products must also be ensured: these are mainly the CleanCap from Trilink (cf. *supra* Part II, Chap. 9, "Solutions for Optimizing the mRNA Molecule"), nucleotide triphosphates and enzymes. These components are not produced by the mRNA companies themselves: they are purchased from specialized biological reagent suppliers (Trilink, and New England Biolabs for enzymes). These vendors have had to increase their production capacity to provide mRNA producers with the necessary materials; this ensured the availability of billions of doses of mRNA vaccines in 2021.

Stability of anti-Covid mRNA vaccines

As mentioned previously, in vitro transcribed mRNA is very stable as a pure (RNase-free) molecule at room temperature. However, the liposomal formulation can be unstable at room temperature and may change (shape, size, percentage of encapsulated mRNA, etc.) due to thawing/freezing. Hence, the special temperature requirements for storage of BioNTech/Pfizer and Moderna vaccines.

Arcturus Therapeutics' self-replicating mRNA vaccine would be less expensive and its manufacturing processes much less demanding. In particular, there would be far fewer cold chain and storage constraints: the need to store BioNTech's anti-Covid mRNA vaccine at − 80 °C and the Moderna vaccine at − 20 °C is a real challenge for developing countries, which often lack the technical means to ensure such storage conditions. As Dr. N. Uddin and Dr. A. Roni point out, inadequate temperature control is currently responsible for wasting half of all vaccine doses. Freeze-dried vaccines have been shown to be much more stable: freeze-drying appears to be a promising way of storing vaccines at temperatures of 2–8 °C, so much so that Pfizer is considering it for its Covid vaccines [2].

Arcturus Therapeutics' vaccine thus represents a promising prospect for the development of a stable vaccine at room temperature, which can only favor a more universal distribution of anti-Covid mRNA vaccines.

Indeed, any improvement in the stability of a vaccine is naturally likely to further facilitate its transport and distribution.

A high-speed train for Pfizer/BioNTech and Moderna, a serious setback for CureVac!

The very rapid commercialization of the BioNTech and Moderna vaccines was made possible by the considerable sums of money committed to increasing industrial capacity.

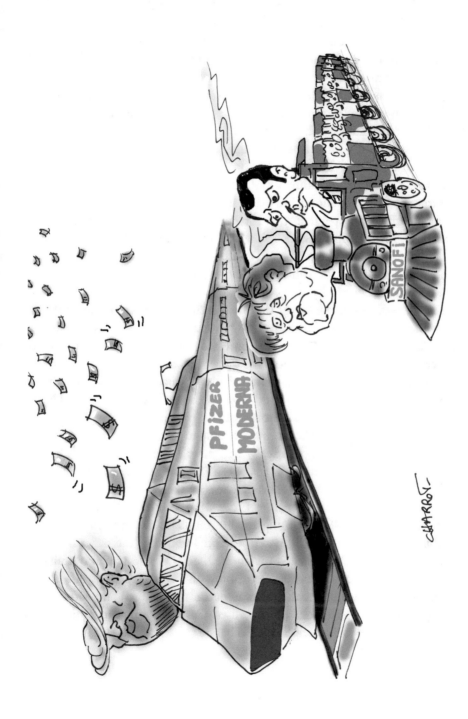

U.S. President Donald Trump's April 2020 launch of Operation *Warp Speed*, to facilitate the development, production, and distribution of Covid vaccines, played a role in their being developed so quickly. Under this program, eight big pharma companies, including Moderna, were selected. They received massive grants from the U.S. government. The budget granted to vaccine manufacturers amounted to a total of 11 billion dollars. Trump did not skimp on spending. It should be noted, however, that the Pfizer/BioNTech partnership, formed on March 17, 2020, did not receive any government funding for the research and production of their vaccine.

Under the terms of the agreement, "Pfizer will pay BioNTech $185 million in upfront payments, including a cash payment of $72 million and an equity investment of $113 million. BioNTech may subsequently receive up to $563 million in milestone payments, for a potential total of $748 million" [3]. Pfizer has provided up to $748 million to make industrial-scale production of the BioNTech vaccine possible. In addition, in July 2020, the U.S. Government placed an order for 100 million doses of the Pfizer/BioNTech vaccine for $1.95 billion [4].

It is the remarkable quality of the BioNTech and Moderna teams that has led to the development of effective mRNA vaccines against Covid in record time. And it is the financial manna injected by the U.S. government, for Moderna, and by Pfizer, for BioNTech, that explains the lead taken by the United States today in the mass production of anti-Covid vaccines and in their distribution to the whole world. Hence, the reaffirmation by the United States of its global *soft power in* the midst of a pandemic!

At the same time, CureVac has been unable to develop an effective mRNA vaccine against Covid, which is really bad news for Europe! Indeed, despite all the expertise accumulated over the past 20 years, the Phase III clinical trial of CureVac's CVnCoV vaccine candidate was marked by a resounding failure, with the biotech reporting, on June 17, 2021, an efficacy rate of its vaccine candidate of only 47%, according to the results of the interim analysis of the Phase III clinical trial. Ultimately, the final results of the same CVnCoV Phase III clinical trial, published on June 29, 2021, indicated an overall efficacy of CureVac's vaccine candidate of only 48%. That's half as effective as the mRNA vaccines marketed by its competitors BioNTech and Moderna!

This was a major setback, which caused CureVac's stock to plummet by nearly 40% on the New York Stock Exchange (Nasdaq) the same day!

The Tübingen biotech has not received government funding comparable to that of Moderna to facilitate the development of its mRNA vaccine candidate. But this cannot explain the disappointing results in the development of the vaccine.

The lack of U.S. funding for the CureVac vaccine is not surprising. Let's recall here the rumored attempt by Donald Trump, in March 2020, to buy the biotech or to convince the company's executives to produce the vaccine exclusively for the U.S. market. This had no other effect than to provoke the anger of the German Chancellor Angela Merkel. CureVac has been working closely with the German government for several years and receives financial support from the Paul-Ehrlich-Institut, a federal drug regulatory agency that reports directly to the Ministry of Health.

Could it be that an insufficient dosage (12 μg for the CureVac vaccine, compared to 30 μg for the BioNTech vaccine and 100 μg for the Moderna vaccine) led to the disappointment in regard to the expected results? In answering that question, it should be noted that these 12 μg of unmodified mRNA contained in the CureVac vaccine induce the expected immune responses (neutralizing antibodies against Sars-CoV-2). So, technically, this vaccine works. It appears that it is the magnitude of this induced immune response that is not sufficient to provide the level of protection demonstrated by the BioNTech and Moderna vaccines.

Perhaps the appearance of the variants also played a role? This explanation has also been put forward by the company's CEO, Franz-Werner Haas [5].

Both of these factors certainly had an impact on the case.

However, it is reasonable to believe that the Tübingen-based biotech may not have surrounded itself with the best expertise in developing its anti-Covid mRNA vaccine, unlike its German competitor BioNTech, whose experts are now widely recognized by the international scientific community.

In any case, this failure of CureVac is probably not related, as demonstrated above (cf. *supra* Part II, "From Preliminary Studies to Clinical Trials"), to a lesser immunogenic action of the unmodified mRNA compared to the modified mRNA. The biotech is currently in clinical trials of a new ("second generation") version of its vaccine, which differs from the first version only in the 3′ stabilization sequences of the mRNA [6]. The rest of the vaccine (liposome, unmodified RNA, and spike coding sequence) is identical to the first version. Therefore, CureVac believes that it was the 3′-stabilization sequence that was insufficient and the key element that accounts for a lower induction of immune response than that seen with the "first generation" BioNTech and Moderna vaccines.

Preclinical data published in *Nature* on November 18, 2021, confirm the very good immune response and effective protection provided by CureVac's second generation mRNA vaccine candidate against Sars-Cov-2 (Cv2CoV vaccine). In particular, preclinical trials in macaques demonstrate an antibody response comparable to that induced by BioNTech/Pfizer's BNT162b2 vaccine [7]. Clinical trials of this vaccine began in March 2022.

At the same time, some *big pharma companies* are working to catch up with the development of mRNA vaccines. This is particularly the case for the French giant Sanofi, which, as of the summer of 2021, had clearly chosen mRNA: after having launched a vast investment program, including the creation of a research center specialized in this technology, the French *big pharma* announced, on August 3, 2021, the acquisition of the American start-up Translate Bio [8], which has recognized expertise, and which had already carried out a clinical trial of mRNA therapy against cystic fibrosis (cf. *supra* Part II, Chap. 14, "Experiments and Clinical Trials Carried Out in Other Therapeutic Fields").

After several months of development of an anti-Covid mRNA vaccine, Sanofi finally announced, on September 28, 2021, that it was halting phase III clinical trials, stating that its vaccine would arrive on the market too late [9].

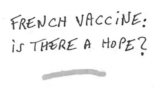

The incidence of variants

Infectiologists agree that the propensity of an RNA virus (including Sars-Cov-2) to mutate is greater than that of a DNA virus.

At the onset of the Sars-Cov-2 pandemic, these experts did not predict a rapid mutation process. But the uncontrolled circulation of products and populations in the world resulted in the infection of a large number of people, with a large number of variants with higher infectious power appearing in the following countries: Great Britain, Brazil, the USA (specifically, the state of California), South Africa, India, Nigeria, France (mainly, Brittany), etc. These mutations were detected in the S protein coding sequence, resulting in an amino acid change and possibly a conformational change of the S protein (i.e., a change in the arrangement of its parts) in its ACE2 receptor binding domain.

In detail, the variants detected in South Africa, Brazil and Great Britain have the same N501Y mutation (replacement of the amino acid asparagine by the amino acid tyrosine). The Brazilian and South African variants also have an E484K mutation, which alters the amino acid sequence in the *Receptor Binding* Domain (RBD) of the S protein—this domain being responsible for recognition and binding with the cell surface receptor ACE2. The Indian variant carries an L452R mutation in the RBD domain that also interacts with the ACE2 receptor.

For the E484K mutation carried by the Brazilian and South African variants, a change in the neutralizing activity of antibodies to Sars-Cov-2 has been observed; however, studies conducted by BioNTech/Pfizer and Moderna in the winter of 2021 were reassuring that this vaccine-induced neutralizing activity is maintained in the presence of these two variants.

With respect to the Indian variant L452R mutation, the *New England Journal of Medicine* published an article in July 2021 [10] that reported that, with BioNTech/Pfizer's BNT162b2 vaccine, two-dose efficacy was 93.7% in those with the alpha variant (commonly referred to as the British variant) and 88.0% in those with the delta variant (commonly referred to as the Indian variant). This demonstrates superior protection compared to Astrazeneca's ChAdOx1 nCoV-19 vaccine, in which two doses were 74.5% effective in people with the alpha variant and 67.0% effective in people with the delta variant. The passage quoted is: "*With the BNT162b2 vaccine, the effectiveness of two doses was 93.7% (95% CI, 91.6–95.3) among persons with the alpha variant and 88.0% (95% CI, 85.3–90.1) among those with the delta variant. With the ChAdOx1 nCoV-19 vaccine, the effectiveness of two doses was 74.5% (95% CI, 68.4–79.4) among persons with the alpha variant and 67.0% (95% CI, 61.3–71.8) among those with the delta variant*".

Would mRNA vaccines based on the S-protein sequence of the original virus be as effective against these variant viruses, which appeared rapidly after the start of the pandemic?

The neutralizing effect of the BioNTech and Moderna mRNA vaccines on the different variants mentioned above—namely, the English, South African, Brazilian and Indian variants—proved to be very good. In particular, the mRNA vaccines were shown to provide effective protection against the most severe forms of Covid 19.

In the fall of 2021, most European countries observed both the emergence of a "fifth wave" of Covid-19, linked to the high contagiousness of the delta variant, and the decrease, over the months, of vaccine immunity, mainly due to a lower antibody response. This double observation led them to progressively generalize the injection of a third dose of the BioNTech or Moderna vaccine.

In December 2021, another variant of Sars-Cov-2 appeared, the omicron variant, which was even more contagious than the previous ones. It soon became clear that the BioNTech and Moderna mRNA vaccines also provided effective protection against the severe forms of the disease caused by this variant. As the Sars-Cov-2 virus evolved, the omicron variant produced a multitude of sub-variants.

Between May and July 2022, most countries in the world experienced a rebound in infections related to the BA. 4 and BA. 5 sub-variants, BioNTech and Moderna adapted their vaccines to make them more effective against the omicron variant. In June 2022, the two biotechs announced conclusive clinical trial results for their so-called "bivalent" vaccines, i.e., demonstrating efficacy against both the original Sars-Cov-2 strain and the omicron variant (including its BA. 4 and BA. 5 sub-variants). In September 2022, the European Medicines Agency issued a positive opinion for the marketing of BioNTech/Pfizer's new anti-Covid mRNA vaccine targeting both the BA. 4 and BA. 5 sub-variants, and Moderna's vaccine targeting these two sub-lineages of the omicron variant was approved on October 21, 2022.

This ultimately demonstrates that, if some variants escape the mRNA vaccines on the market, the biotechs that develop them will respond very quickly by designing new vaccines that would accommodate the mutation. Recall that Moderna produced its mRNA vaccine in just 42 days, from January 13 to February 24, 2020. In the near future, the timelines could be even shorter: reasonably, one month to produce, and three to four months to test (if necessary, depending on the choice of health authorities). The marketing authorization could benefit from an accelerated procedure.

Considering the likely scenario of Covid-19 becoming a seasonal infection with the presence of infectious variants, like influenza, the prophylactic response could consist in combining several mRNA vaccines in a single injection.

This should not, of course, lead to the conclusion that vaccination is useless for the treatment of Covid-19, because, in fact, mRNA vaccines have contributed to saving millions of lives by providing immune protection to entire populations, and by helping to reduce the severity of the most severe cases in patients. In addition, they offer new opportunities for personalized care. In this sense, the mRNA vaccine designed to treat Covid-19 is not the end of the story, but rather the beginning of a medical adventure whose promises, outlined as early as the 1990s, will be fully realized in the years to come. In this case, since flu vaccines could also be made with mRNA, it would be possible to combine several vaccines in a single injection.

References

1 Steve Pascolo, "Emerging from the Pandemic", article published on February 17, 2021 on the Swiss Medical Forum, and available at: https://medicalforum. ch/fr/detail/doi/fms.2021.08742.

2 Mohammad N Uddin, Monzurul A Roni, "*Storage and stability challenges of COVID-19 mRNA-based vaccines*", in *Vaccines*, September 17, 2021, doi: https://doi.org/10.3390/vaccines9091033.

3 BioNTech/Pfizer press release of April 9, 2020, "*Pfizer and BioNTech unveil new details about their collaboration to accelerate the global development of a COVID-19 vaccine*", CP available on Pfizer's website at https://www.pfizer.fr/ pfizer-et-biontech-devoilent-de-nouveaux-details-propos-de-leur-collaboration-en-vue-daccelerer-le.

4 Catherine Ducruet, "*Coronavirus: why Trump is ordering vaccines in the hundreds of millions*", Les Échos, July 22, 2020. Article available *at:* https://www.lesechos. fr/industrie-services/pharmacie-sante/coronavirus-pourquoi-donald-trump-com mande-des-vaccins-par-centaines-de-millions-1225615.

5 "*Europe's stock of the day. CureVac: variants play a bad trick on the vaccine*", in *Capital*, June 17, 2021. Agence Option Finance (AOF) article available at: https://www.capital.fr/entreprises-marches/la-valeur-du-jour-en-europe-curevac-les-variants-jouent-un-mauvais-tour-au-vaccin-1406769.

6 CureVac Press Release, August 16, 2001, "*Second-Generation mRNA COVID-19 Vaccine Candidate, CV2CoV, Demonstrates Improved Immune Response and Protection in Preclinical Study*", PC available at: https://www.curevac.com/en/ 2021/08/16/second-generation-mrna-covid-19-vaccine-candidate-cv2cov-dem onstrates-improved-immune-response-and-protection-in-preclinical-study/.

7 Makda S. Gebre, M.S., Rauch, S., Roth, N. G Nicole Roth, Jingyou Yu, Abishek Chandrashekar, Noe B. Mercado, Xuan He, Jinyan Liu, Katherine McMahan, Amanda Martinot, David R. Martinez, Tori Giffin, David Hope, Shivani Patel, Daniel Sellers, Owen Sanborn, Julia Barrett, Xiaowen Liu,

Andrew C. Cole, Laurent Pessaint, Daniel Valentin, Zack Flinchbaugh, Jake Yalley-Ogunro, Jeanne Muench, Renita Brown, Anthony Cook, Elyse Teow, Hanne Andersen, Mark G. Lewis, Adrianus C. M. Boon, Ralph S. Baric, Stefan O. Mueller, Benjamin Petsch & Dan H. Barouch, "*Optimization of Non-Coding Regions for a Non-Modified mRNA COVID-19 Vaccine*", *Nature*, 18 November 2021. DOI: https://doi.org/10.1038/s41586-021-04231-6.

8 *Le Monde*, "*Sanofi buys messenger RNA specialist Translate, for $3.2 billion*", August 3, 2021. Article available on the Le *Monde* website at: https://www.lem onde.fr/economie/article/2021/08/03/sanofi-rachete-le-specialiste-de-l-arn-mes sager-translate-pour-3-2-milliards-de-dollars_6090379_3234.html.

9 Le *Monde*, "Covid-19: Sanofi halts development of its messenger RNA vaccine, but continues its most advanced project", September 28, 2021. Article available online at: https://www.lemonde.fr/planete/article/2021/09/28/covid-19-sanofi-arrete-le-developpement-de-son-vaccin-a-arn-messager_6096260_3244.html.

10 Jamie Lopez Bernal, Nick Andrews, Charlotte Gower, Eileen Gallagher, Ruth Simmons, Simon Thelwall, Julia Stowe, Elise Tessier, Natalie Groves, Gavin Dabrera, Richard Myers, Colin N.J. Campbell, Gayatri Amirthalingam, Matt Edmunds, Maria Zambon, Kevin E. Brown, Susan Hopkins, Meera Chand, Mary Ramsay, "*Effectiveness of Covid-19 Vaccines against the B.1.617.2 (Delta) Variant*", in *New England Journal of Medicine*, July 21, 2021, vol. 385, p. 585–594. DOI: https://doi.org/10.1056/NEJMoa2108891.

20

mRNA, a Technology of the Future

The triumph of the anti-Covid vaccine opens the door to other medical applications for mRNA. Future breakthroughs depend on the development not only of biotechnologies, but also of biotechs, particularly in Europe.

We can mention

i. The Development of Nanotechnology

The use of alternative mRNA vectors to liposomes could optimize transport to the target cell, while reducing the dose of mRNA injected into the patient.

These vectors include:

- **polymers**, the binding of amino acids to positively charged polymers results in the formation of "polyplexes", candidate mRNA vectors, which are currently under study. Although the results of clinical trials of these polymeric compounds do not match liposomes today, there is significant hope: the technology is progressing, and the "polyplexes" are very stable. Within a few years, they will be promising candidates for mRNA vaccination;
- **protamines**, already proposed in the founding works of CureVac, are still relevant;
- **dendrimers**, hydrophilic macromolecules with a tree-like structure, able to bind to membranes and to be encapsulated, can transport nucleic acids.

© The Author(s), under exclusive license to Springer Nature Switzerland AG 2023
J. Lemonnier and N. Lemonnier, *The Marathon of the Messenger*,
https://doi.org/10.1007/978-3-031-39300-6_20

Biocompatible, they are not immunogenic. Their industrial production is under study;
- **hybrid systems**, which combine the main existing vectors for mRNA transport, are also possible. These are mainly the lipopolyplexes mentioned above (cf. *supra* Part II, Chap. 9, "Solutions for mRNA Optimization"), developed by Patrick Midoux and Chantal Pichon, in collaboration with EtheRNA.

ii. Monoclonal Antibody Production via mRNA

In investing in these fast-growing drugs, the pharmaceutical industry is looking for simple and fast production alternatives that are also cheaper. Produced in small-scale biotech, mRNA technology is a choice that requires improvements to reach industrial quantities open to clinical trials.

Thus, in this particular field, as stated by Itziar Gómez-Aguado et al.: "*passive immunization by mRNA encoding monoclonal antibodies is showing great biomedical interest. Given the rapidly growing market of therapeutic mono-clonal antibodies, and the high cost of this type of medicines, the pharmaceutical industry is looking for alternative approaches. mRNA is considered a good option, due to its simpler, faster and more cost-effective synthesis. Until now, pre-clinical studies in small rodents have demonstrated antibody titters from the first day after mRNA intravenous administration. However, before moving on to the clinic, it is still necessary to highlight whether mRNA can lead to high antibody concentrations, often needed for a therapeutic effect*". [1]

iii. Chemical Synthesis of mRNA

Another possibility is the direct synthesis of small mRNAs (on the order of 100 ribonucleotides). In this field of biotechnology, current studies are investigating the impact of chemical modifications to the mRNA on protein translation. Indeed, linkage errors are more frequent than with the use of bacterial plasmids (as a reminder, DNA production, then in vitro transcription into mRNA). The progress made is significant, but many steps remain to be taken before chemical synthesis of mRNA, a promising technological option, can be achieved. In particular, Steve Pascolo, at the University of Zurich, has recently succeeded in developing fully chemically synthesized mRNA molecules [2].

(i) Biotech mRNA: European and French developments

- *European developments*

The European Union wants to promote research programs that use nano-materials. In this case, it wants to accelerate the development of mRNA technologies for the immunotherapy of breast cancer and the treatment of damaged heart tissue (tissue regeneration).

The so-called EXPERT (*EXpanding Platform for Efficacious mRNA Therapeutics*), launched in 2019 as part of the European Union's Horizon 2020 innovation program, is funded by the EU with €14.9 million. Its primary objective is to materialize collaborations between companies and academic partners from eleven different countries (these are the Netherlands, Belgium, Norway, Sweden, Spain, Hungary, Ireland, France, Germany and Israel).

Within the European Union, the European Research Council commissions currently support research programs through grants awarded according to the classic process: *calls for grants, and evaluation of projects by independent experts appointed by the commissions' scientific councils.*

> The *European Research Council* is a body of the European Union responsible for coordinating the research efforts of the various Member States. It offers funding for European research projects through grants.

For the years 2014–2020, the *grant* funding was part of Horizon 2020, which was, during this period, the European Union's support program for research and development; it is part of Horizon Europe for the period 2021–2027. The *ERC Starting Grant* provides support to young researchers with promising scientific careers to build interactive research teams at the European level. A *Call for grant* has been launched on mRNA, to finance innovation and industrialization projects.

- *French developments*

In the fall of 2020, aware that France was lagging behind in the field of biotherapies and biotechnologies, the French government wanted to encourage the development of a strategy involving all industrial players, through a global think tank set up a few years earlier, the Comité stratégique de filière des industries et technologies de la filière Santé.

On December 14, 2020, this committee announced the creation of a scientific and industrial steering structure for the entire sector, the "Alliance France

Bioproduction", whose objective is to mobilize the skills and resources available to make France the European leader in pharmaceutical biomanufacturing by 2030 [3].

Thirteen months later, on January 7, 2022, and as part of the new industrial development policy driven by French President Emmanuel Macron, France 2030, the French government officially launched the "Biotherapies and Bioproduction Acceleration Strategy for Innovative Therapies" [4]. This is a global strategy that integrates various public support mechanisms. Its stated objective is to strengthen France's competitiveness, position and attractiveness in the field of biotechnology, with a view to leadership by 2030. To achieve this, the measures implemented must encourage the industrial and academic players concerned to invest in research, innovation and industrial development.

The sought-after development of mRNA therapies in France is a natural part of this strategy.

With the health crisis, the European pharmaceutical industry, and the French industry in particular, has become aware of the major interest in mRNA for a certain number of therapies, such as the treatment by vaccination of cancers and infectious or emerging diseases. In the next few years, mRNA will cease to be subject to exceptional usage, such as during a pandemic, and will become much more common in its usage: it could then become a new class of drugs marketed in France. Eventually, the consequences on the patient's pathway to health will also have to be taken into account.

References

1 Itziar Gómez-Aguado, Julen Rodríguez-Castejón, Mónica Vicente-Pascual, Alicia Rodríguez-Gascón, María Ángeles Solinís, Ana del Pozo-Rodríguez, *"Nanomedicines to Deliver mRNA: State of the Art and Future Perspectives, "Nanomaterials"*, 20 February 2020. DOI: https://doi.org/10.3390/nano10 020364.

2 Steve Pascolo, *"Minimal messenger RNAs and their use"*, Swiss patent (University of Zürich), No. WO2021038089, international filing date: August 28, 2020, publication date: March 4, 2021. Available at: https://patentscope.wipo.int/sea rch/fr/detail.jsf?docId=WO2021038089&_cid=P20-KNA37Y-85597-1.

3 *"Creation of the Alliance France Bioproduction: making France the European leader in bioproduction by 2030"*, article published on the CEA website on December 14, 2020, available on the internet at the following address: https://www.cea.fr/presse/Pages/actualites-communiques/sante-sciences-du-vivant/alliance-france-bioproduction.aspx.

4 "*A strategy for a leading France in the production of innovative therapies*", on the France 2030 portal of the Ministry of the Economy. Article published online on January 10, 2022 and available at the following address: https://www.economie.gouv.fr/strategie-france-leader-production-therapies-innovantes.

Conclusion

We are nothing. What we seek is everything.

Friedrich Hölderlin.

On May 13, 1961, two original papers appeared in the same journal, *Nature,* the result of different experiments carried out in two different American laboratories. These two articles marked the true birth of messenger RNA. Two French researchers from the Institut Pasteur in Paris were closely associated with this discovery: François Gros, first author of one of the papers [1], and François Jacob, second author of the other [2].

© The Editor(s) (if applicable) and The Author(s), under exclusive
license to Springer Nature Switzerland AG 2023
J. Lemonnier and N. Lemonnier, *The Marathon of the Messenger,*
https://doi.org/10.1007/978-3-031-39300-6

"[...] two original papers appeared in the same journal Nature."

From the discovery of mRNA to its approval for therapeutic use, it took 60 years!

It was a real marathon launched in the last decade of the twentieth century, during which a very small number of adventurers, who could literally be counted on the fingers of one hand, had committed themselves to the mRNA cause, and faced the winds of general skepticism, without any support. The first decade of the twenty-first century saw a few others join these reckless people. The concept of injecting synthetic RNA was still far from being accepted by all immunologists: mRNA was expensive, its stability was in question, and the results appeared uncertain for a long time. During the second decade, the accumulation of significant results suddenly made the field grow: some were moved by their faith in a new therapy, others by opportunism…

Today, with the triumph of mRNA against Covid-19 and the opening of vaccine therapy for other diseases, four elements deserve to be highlighted.

1. **Contrary to the messages spread during the winter of 2021, the development of mRNA vaccines is primarily a European story.**

First of all, it is important to remember the key role played by German and French researchers: Pierre Meulien, Hans-Georg Rammensee, Ingmar Hoerr, Steve Pascolo, Ugur Sahin and his entire team at BioNTech at the beginning of the 2010s. Since then, CureVac, in collaboration with the University of Tübingen and the University Hospital of Tübingen, has broken new ground and initiated human clinical trials of mRNA cancer vaccines. After 2010, by improving the "packaging" of the mRNA, BioNTech generated the diffusion of a vaccine therapy into medical practice.

With the success of the Covid vaccines, many media outlets have focused on the role of Drew Weissman and Katalin Kariko in the development of a modified "de-immunized" mRNA, through the replacement of uracil with pseudouridine. Originally, this mRNA modification was supposed to avoid the induction of an immune reaction. However, in fact, this modified mRNA was *ultimately* used by BioNTech and Moderna to trigger an immune response against Covid-19.

In this case, however, it appears entirely possible to use unmodified mRNA for vaccination without inducing an uncontrolled immune response. This has been amply demonstrated by the work of CureVac and BioNTech. More recently, the phase-III clinical success of Arcturus Therapeutics, based on self-replicating mRNA, also demonstrates that mRNA modification is not the be-all end-all of messenger RNA vaccination.

And the diversity of therapeutic applications of messenger RNA is becoming increasingly apparent, as evidenced by the growing number of clinical trials in many countries around the world. It is no longer only French and German researchers, but also American, Chinese, Canadian, Swiss, English, Dutch, Indian, etc., who are using it today to open up new avenues for the medicine of the future.

2. Risk plays a very important role in the scientific process

The first clinical trials in 2003 against melanoma, and the first injection of synthetic mRNA produced under GMP conditions published in 2007 to express a protein of interest, represented a leap into the unknown. These unprecedented experiments paved the way for a new therapy.

It was a real gamble for the scientists who made it. Let us admire the courage and enthusiasm of these researchers, who, in doing so, put their scientific credibility, as well as the capital of the biotech company at which they worked, on the line! This represented a considerable risk for them; but this risk was, and still is, the essential condition to make possible extraordinary progress in biology and medicine.

Napoleon 1st stated: "*He who knows where he is going, does not usually go very far*". Randomized trials are always pre-determined to achieve a targeted objective: they confirm or validate the merits of a drug or the conditions of its use. In public health, the real heuristic approach is that of the "front runners", who, starting from fundamental science, imagine and design a new way to propose a new therapy. They have the courage to challenge the unknown, motivated by the altruistic need to advance humanity by fulfilling themselves. When mRNA aroused the interest of a few researchers, the therapeutic prospects remained uncertain. They saw the risk as an exciting and fantastic opportunity; more than that, it can be said that the prevailing skepticism of the scientific community stimulated them.

The clinical trials for the Covid vaccines developed by BioNTech and Moderna showed that the risk does not just belong to the researcher. It was also carried by the tens of thousands of people who received injections of products whose potential impact on the body was not known at the time. The risk is indeed societal. One can only admire these voluntary guinea pigs, unknown but essential actors for scientific progress; thanks to them, vaccines have been developed that have saved millions of lives. This is the same challenge that Pasteur faced when he developed the rabies vaccine. Thanks to these courageous and anonymous people, we now benefit from the mRNA vaccine in complete safety.

3. Financial support for biotechs, key players in the health sector

The knowledge in biology extracted from living organisms by academic researchers constitutes mines to be exploited for human health. Biotechs are the only companies capable of making this transfer. As bridges between the academic world and industrial production, they transform the cognitive into a treatment!

A start-up can benefit from governmental aid at its beginning, when it is coupled with a specific structure of accompaniment of the company's projects of creation called an incubator. However, this aid is temporary. The pharmaceutical industry can also support the company on an ad hoc basis, or even partner with it for development or clinical trials, but essentially only in the case of a solid project, with a short-term completion date, and the certainty of a significant return on investment (*Return Of Investment,* "ROI"). In fact, it is up to the biotech to assume this risk, which is always high. Through the quality of the project presented, it must convince investors to take the risk of winning or losing! This reality concerns not only a new technology (mRNA vaccination), but also the repositioning of old molecules.

Boldness and entrepreneurial freedom lead to the creation and emergence of new technologies, at the cost of significant investments and sometimes even personal risks: for example, the founders of CureVac have experienced several years of very low salaries. The triumph of mRNA is a perfect example of the roles played and accomplishments achieved by researchers and actors –academics (from universities and higher education institutions), biotechs, and private industry. For the continuation of a sustainable interactive model, it is essential that each party finds its place and participates in the success.

4. The moral values associated with scientific progress, namely, intellectual curiosity and the will to innovate, still have meaning in our time.

The will to innovate is at the very heart of the scientific process. Of course, we must be aware of the potential consequences of scientific discoveries and be particularly vigilant on this subject. These fears are legitimate at a time when biological engineering can give humanity a false impression of omnipotence.

However, in the case of mRNA, which is quickly degraded and cannot be integrated into the human genome, there is no reason to fear genetic manipulation. No one can intervene to create a monster; let Mary Shelley's *Doctor Frankenstein* rest in peace!

"There is no reason to fear genetic manipulation."

The precautionary principle, which is generally put forward, must not have the effect of discouraging curiosity and the taste for innovation, because these are the real drivers of progress in biology.

This requires the development of critical thinking, scientific rigor and open-mindedness. These qualities are particularly valuable at a time when so much erroneous information is circulating, taken up without real examination or analysis by social networks, and even by the media. Only objective knowledge can help to form a clear and realistic idea of things, and not a reliance on "mythological" thinking, which, in the end, proves to be destructive and misleading. The dissemination of accurate information on medicine responds to this concern for objectivity; *in fine,* it is the indispensable condition for a good understanding of the therapeutic progress that messenger RNA technology makes possible today.

1. François Gros, H. Hiatt., W. Gilbert, Kurland., R.W. Risebrough, & James D. Watson, "*Unstable Ribonucleic acid Revealed by Pulse Labelling of Escherichia coli*", *Nature*, 13 May 1961, vol. 190, p. 581–585. DOI: https://doi.org/10.1038/190581a0.
2. S. Brenner, François Jacob & M. Meselson "*An Unstable Intermediate Carrying Information from Genes to Ribosomes for Protein Synthesis*", *Nature*, 13 May 1961, vol. 190, p. 576–581. DOI: https://doi.org/10.1038/190576a0.

mRNA INDUCTION OF INTERFERON

1963 :
A. Isaacs
R. A. Cox
Z. Rotem.

NAKED mRNA INJECTION

Protein translation in vivo
(in mice) after injection of naked
mRNA injection.
1990 :
J. A. Wolff, R. W. Malone.

– The Premises –

1961

Credit : Institut Pasteur - 58262

NOBEL PRICE 1965

mRNA Discovery :
François Jacob
Jacques Monod
and André Lwoff.

DEVELOPMENT 1st LIPOSOMES ENCAPSULATION mRNA
Nature 274, 923–924
1978 : G. J. Dimitriadis.

**FIRST mRNA VACCINE
(ON MICE)**

First worldwide vaccine trial
on mice (influenza).
1993 :
F. Martinon, P. Meulien,
G. Lenzen, S. Krishnan.

**FIRST mRNA VACCINE
(HUMANS) AND 1st
CLINICAL TRIAL**

1st injection of mRNA in humans
(S. Pascolo);
melanoma clinical trial
2003 : B. Weide, S. Pascolo.

Institut Pasteur

– The pioneers –

1992   **2003**   **2005**   **2010**

**NEW mRNA VACCINE
TESTS ON MICE**

Ingmar Hoerr, HG Rammensee.

UNIVERSITAT
TÜBINGEN

CUREVAC
the RNA people®

MODIFIED mRNA

Reduction of the immunogenic
character of the mRNA, and
increase in the number of
proteins produced
(opens the way to other therapies)
2005 : K. Kariko, D. Weissman.

CLINICAL TRIALS CANCER

First clinical trials
of BioNTech mRNA on cancer
2010 : U. Sahin, Ö. Türeci.

– Validation and clinical tests –

2005   2010

CleanCap

FIRST "BIG-PHARMA" AGREEMENTS

 Pfizer   MSD   AstraZeneca

**1ST USE mRNA
MODIFIED FOR VACCINE**

– Pandemic Covid-19 –

2020

PANDEMIC Covid-19

Jan 21: genetic code of Sras-CoV-2 published.
Mar 21: Moderna mRNA vaccine ready for testing.

(c) BioNTech (Wolfgang Wilde)

Pfizer

moderna

(c) The Messenger's Marathon

Postfaces

Afterword by Chantal Pichon

This story shows us that science requires intelligence, passion, patience, vision and luck.

Thanks to the success of these mRNA vaccines, biotherapies and nanomedicine have gained real interest. These vaccines would not have succeeded without the interaction of several disciplines. However, while such interdisciplinarity is often claimed, it is rarely supported.

Let's hope that this success will encourage the funding of interdisciplinary research programs; above all, we must avoid a scattering of funds that would make real technological leaps that much harder. There is no doubt about it: we are at the dawn of a new revolutionary era in medicine, particularly in the fight against cancer.

mRNA offers us the possibility of both personalized therapy—by identifying the predominant neoantigens in a patient—and more universal strategies to turn a so-called "cold" tumor into a "hot" tumor in order to thwart its inhibitory action and help the immune system respond.

How many innovative strategies and technologies that deserve to be developed are lurking in the shadows? To bring them to the fore, we need to break down the silos to get a clear picture of all the research underway, and have the courage to fund high-risk projects, some of which are sure to be sources of change and innovation.

Chantal Pichon

J. Lemonnier and N. Lemonnier, *The Marathon of the Messenger*,
https://doi.org/10.1007/978-3-031-39300-6

Afterword by Steve Pascolo

Messenger RNA is present in all life forms. It codes for proteins, and thus generates all the functions of living organisms. Since the 80s, we have known how to produce it in vitro. Despite its pluripotency, until 2019, synthetic messenger RNA was generally despised by the pharmaceutical industry and applied research, because it was perceived as being too fragile to be used as a basis for drugs. However, the natural and rapid degradation of messenger RNA in the body is actually a guarantee of safety in the context of therapeutic or vaccine applications of this molecule in its synthetic form. Only a very small number of scientists, mostly in Germany, have believed in messenger RNA as a drug, and they have been developing it in the shadows since the 1990s. It is this long-term work that made it possible to validate the Covid-19 vaccines in less than a year in the year 2020.

The synthetic messenger RNA vaccine is essentially a French story and a German story (a true success story in the German case). It took 30 years of research, from the first publication of the synthetic messenger RNA vaccine by the French in 1993, to the first clinical studies in Germany in 2003, to the approval of a synthetic messenger RNA vaccine against Covid-19 by the German company BioNTech. Numerous improvements in the technology, mainly by biotech companies, have transformed the rudimentary vaccines developed by the synthetic messenger RNA pioneers in the 1990s into highly effective and still safe vaccines, due to the nature of the messenger RNA, which is destroyed very quickly after injection.

The influence of synthetic messenger RNA technology is not limited to Covid-19 vaccines, but extends to all areas of preventive and therapeutic medicine. In theory, it is possible to define a therapeutic strategy with synthetic messenger RNA for each disease. After vaccines to prevent infectious diseases, we will probably finally see the completion of anti-cancer vaccines upon which the pioneers of messenger RNA have worked; we will certainly also see the appearance of therapeutic proposals to treat degenerative diseases (e.g., Alzheimer's or Parkinson's), autoimmune diseases (multiple sclerosis, type 1 diabetes, etc.) or genetic diseases (cystic fibrosis, myopathies, etc.), or even to offer methods of regeneration or to manage aging. The therapeutic possibilities of synthetic messenger RNA are immense, and with the approval of BioNTech/Pfizer and Moderna's Covid-19 vaccines, a new therapeutic era is beginning.

Steve Pascolo

"Most of the mRNA vaccine research was conducted in Europe."

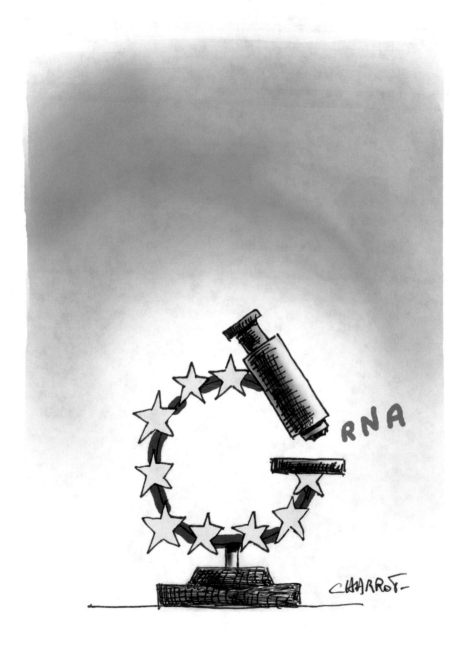

Glossary

Glossary of Common Names

Adjuvant therapy in cancer, a treatment that complements a primary treatment, with the goal of increasing its effectiveness. For example, chemotherapy can be an adjuvant treatment to surgery or radiation therapy.

Amino acid organic molecule with a carbon skeleton that has an amine group and a carboxyl group; it is the basic component of all proteins.

Antigen substance recognized by adaptive immunity (antibodies or T cells).

Antigen-presenting cells (APCs) cells of the immune system that present parts of intruding elements to T cells: dendritic cells, possibly monocytes, macrophages, and B cells.

Apoptosis physiological process of programmed cell death.

Autologous refers to a tissue or cell transplant when the donor and the recipient are the same individual.

B lymphocyte a particular white blood cell that is part of the lymphocytes; cells synthesized in the bone marrow that circulate in the blood and lymph to participate in the natural defenses of the body.

Big Pharma the pharmaceutical industry as a whole and, above all, the largest groups on a global scale.

Biotech a company that develops and produces drug substances by applying scientific and technological discoveries to living organisms or their components.

CAR-T Cells a patient's own T cells that have been genetically modified to recognize and destroy cancer cells.

J. Lemonnier and N. Lemonnier, *The Marathon of the Messenger*, https://doi.org/10.1007/978-3-031-39300-6

Codon triplet of nucleotides present on the mRNA of a given gene that specifies the tRNA in charge of presenting one of the 20 amino acids necessary for the construction of the protein encoded by this gene.

Covid-19 *corona virus disease*, a respiratory disease discovered in 2019, and transmitted by a coronavirus, Sars-CoV-2; it is responsible for a still-active (as of this writing) global epidemic—the origin of which is still debated—that emerged in December 2019 in the city of Wuhan, Hubei province, China. It is potentially fatal in subjects weakened by age or other chronic disease.

CRISPR-CAS9 a biological tool that cuts the DNA of a cell at a specific sequence, consisting of a guide RNA (CRISPR) that targets the sequence of interest and the enzyme CRISPR associated protein 9 (CAS9), the "molecular scissors" of DNA.

Cytokines a group of polypeptide molecules (interleukins and interferons), produced by the immune system, that engage in various activities, all of which regulate the body's immune defense. Among the cytokines, **chemokines** are small proteins that direct the movement of white blood cells under the influence of various chemical substances.

Cytotoxic T lymphocyte a killer cell essential to the immune response against viruses and cancers and characterized by the expression of the CD8 molecule (T8 cells).

Dendrimer a large, tree-shaped polymer produced by iterative construction and endowed with active sites; drug delivery material.

Dendritic (cell) "sentinel" cell of the innate immune system, presenting antigen of the "non-self" that triggers the adaptive response.

DNA deoxyribonucleic acid.

DNA replication a process that duplicates the initial sequence, to create a molecule identical to the base molecule.

Endosome cytoplasmic vesicle formed by the folding of a lipid membrane.

Epitope part of an antigen recognized by a receptor located on the surface of a lymphocyte or by an antibody.

Epizootic an epidemic that strikes animals. *See also* **Zoonosis**.

Erythropoietin (EPO) a hormone that stimulates the production of erythrocytes (red blood cells).

Facilitating antibodies antibodies that promote viral infection of cells that carry the antibody Fc receptors (or immunoglobulins).

GMP shorthand for 'good manufacturing practice' in regard to medicines that meet the requirements of a national or international health system.

Grant a grant or scholarship awarded by a state, an organization, or a foundation to a research project, following a call for proposals and an evaluation by a scientific council.

Immune reaction defense mechanism activated by the organism in response to the identification of a pathogen intrusion, an aggression or a dysfunction of the organism in question.

Immunogenicity the ability of an antigen to provoke a specific immune response.

Impact Factor an indicator that estimates the visibility of a scientific journal, and a quantitative evaluation tool for research.

Inflammatory reaction response of living and vascularized tissues to an aggression; clinically manifested in a variable way, most often with fever and alteration of the general state.

Intellectual property protection and promotion of inventions and innovations. *See also* **Patents.**

Interferon polypeptide cell secretion after viral infection, which inhibits the production of viruses of different species.

Key Opinion Leader a person or organization recognized and respected by the public for their knowledge and expertise in a given field.

License to operate granting of a third party (the licensee) the right to use a patent and the technical processes described therein.

Liposome microscopic artificial vesicle capable of introducing a substance into the cells of an organism.

Luciferase enzyme that produces light from luciferin molecules.

Major histocompatibility complex (MHC) a group of molecules used to recognize the "self", involved in the phenomena of transplant rejection and recognized by T cells.

Marketing authorization (MA) national or European authorization issued, after evaluation, to the holder of a pharmaceutical company responsible for marketing a drug.

Messenger RNA (mRNA) a copy of a region of DNA that represents a gene; synthesized in the cell nucleus, mRNA transmits information about the manufacturing plan of the protein encoded by this gene.

Modified mRNA synthetic mRNA derived from the transcription of the DNA sequence of the target gene, in which the uracil base is replaced with pseudo-uridine *(see this word)*.

Monoclonal antibodies antibodies produced by a homogeneous line (a clone) of lymphocytes that recognize a single epitope on a single antigen.

Nanotechnology all studies and processes for the manufacture and manipulation of structures, devices and material systems on the nanometric scale (the order of magnitude of the distance between two atoms).

Nucleoside basic element of the nucleic acids, made up of a nitrogenous base associated with a sugar (ribose for the RNA, and deoxyribose for the DNA).

Nucleotide linkage of a nucleoside to one to three phosphate groups.

Opsonization process by which a pathogen (bacteria, a cell infected by a pathogen…) is marked by a molecule called opsonin for ingestion and destruction by phagocytes equipped with receptors for opsonins.

Patent industrial property right granted by a public authority (INPI, in France) to the person who discloses, describes in a complete and sufficient manner and claims an invention, in order to benefit from a monopoly of exploitation on the latter, and to protect it against possible unauthorized copies (or counterfeits). *See also* **Intellectual Property.**

Pattern Recognition Receptors (PRRs) receptors that recognize the molecular patterns of pathogens; specific to the innate immune system.

Polypeptide chain of 10–100 amino acids linked by peptide bonds.

Prophylaxis a set of measures to prevent the deleterious action of an infectious agent (bacteria or virus). *See also* **Vaccination**.

Protein Replacement a therapy that provides a missing protein to an organism.

Pseudo-uridine isomer of uridine; these are **isomers of** compounds that have the same overall chemical formula, but different properties, because of a disctinct arrangement of atoms in the molecule.

Regenerative medicine medicine that repairs the function of a damaged tissue or organ.

Reviewer an expert in a specific field mandated by the editorial board of a scientific journal to evaluate an article submitted to that journal.

Ribosome a cellular organism that translates messenger RNA into proteins, a kind of "translation factory" for the genetic code.

RNA ribonucleic acid.

Sars-CoV-1 a linear RNA virus of the *coronavirus* genus, responsible for a form of severe acute respiratory syndrome (SARS) that occurred between 2002 and 2004.

Sars-CoV-2 virus responsible for Covid-19. *See also* **Covid-19**.

Spike (protein) a constitutive protein of the envelope spicules of Sars-CoV-2, which is one of the targets of the immune system in the face of infection, and of current mRNA vaccines.

Spin-off young company resulting from the spin-off of a larger one.

T helper lymphocyte original type of T lymphocytes, non-cytotoxic, at the center of the immune response, and characterized by the expression of the CD4 molecule (T4 cells).

Toll-like receptor a type of pattern recognition receptor (PRR) that plays a key role in the adaptive immune system by alerting it to the presence of microbial infections.

Transcription biological mechanism that leads to the synthesis of an RNA molecule from a DNA molecule.

Transdifferentiation the ability of already differentiated cells to lose their normal characteristics and acquire new ones, along with new functions.

Transfection introduction of a foreign nucleotide sequence into a host cell.

Transfer RNA (tRNA) a short RNA that specifically contributes an amino acid to the machinery of protein production.

Translation the process of synthesizing a polypeptide chain or protein from a messenger RNA.

Unmodified mRNA synthetic—native—mRNA derived from the transcription of the DNA sequence of the target gene.

Vaccination injection of an infectious agent (virus or bacteria), in a harmless form, or one of its components, into the body to stimulate the immune response. *See also* **Prophylaxis**.

Variant of a virus virus that has one or more genetic mutations in relation to its original form.

Zoonosis a disease or infection of vertebrate animals that can be transmitted to humans, and vice versa. *See also* **Epizootic**.

Glossary of Proper Names

BioNTech German biotechnology company based in Mainz, specializing in the development and production of immunotherapeutics active in the treatment of serious diseases.

CureVac German biotech company, headquartered in Tübingen, that develops therapies using native synthetic mRNA to treat infectious diseases, cancer and rare diseases.

Gros, François (1925–2022) French biologist, one of the pioneers of French cellular biochemistry, and co-discoverer of mRNA; director of the Institut Pasteur in Paris (1976–1982), honorary professor at the Collège de France, and member of the Institut de France.

Hoerr, Ingmar (born 1968) German biologist, and one of the founders of CureVac. He played an important role in the early years of the biotech company by contributing to its economic and commercial development.

Jacob, François (1920–2013) French biologist and physician; discoverer of mRNA; winner of the Nobel Prize in Medicine (1965); professor at the Institut Pasteur in Paris and at the Collège de France, member of numerous foreign academies, and companion of the Libération.

Kariko, Katalin (born 1955) Hungarian-born biochemist working in Pennsylvania, specializing in mRNA biotechnology; former Vice President of BioNTech RNA pharmaceuticals (until October 2022).

Meulien, Pierre Irish researcher, who first demonstrated, in 1993, that mRNA works as a vaccine; currently executive director of the *Innovative Medicines Initiative*, a public–private partnership between the European Union and major European pharmaceutical companies.

Moderna, Inc U.S. biotech company, originally Mode RNA, founded in 2010, based in Cambridge, MA (USA), focused on protein therapies based on messenger RNA technology.

Pascolo, Steve (born 1970) French researcher, biologist and immunologist, and one of the pioneers of mRNA vaccine technology, especially against cancer; co-founder and scientific director of CureVac (2000–2006), director of an mRNA platform for cancer research (University Hospital Zurich), and managing director of the Swiss company Miescher Pharma GmbH, Zurich.

Pasteur (Institute) French private non-profit foundation dedicated to the study of biology, microorganisms, diseases and vaccines. Created in 1888, thanks to an international public subscription, and named after the famous scientist Louis Pasteur. *See also* **Pasteur, Louis**.

Pasteur, Louis (1822–1895) French chemist, physicist and microbiologist, founder and first director of the eponymous institute, which, among many other discoveries, developed the first rabies vaccine in 1885. *See also* **Pasteur (Institute)**.

Pfizer U.S. pharmaceutical company founded in 1849, and world leader in the field of pharmacy.

Pichon, Chantal university professor, French researcher, and specialist in molecular and cellular biology. She has developed therapeutic strategies based on nucleic acids, and has been conducting research on messenger RNA technologies since 2005.

Rammensee, Hans-Georg (born 1953) German immunologist and professor at the University of Tübingen, he worked on the development of cancer immunotherapies. He joined CureVac a few months after its foundation in 2000.

Sahin, Ugur (born 1965) German physician and entrepreneur, President and CEO of BioNTech. A renowned oncologist, he and his wife, Özlem Türeci, founded Ganymed Pharmaceuticals in 2001 and BioNTech in 2008. He developed the first vaccine against Covid to be marketed in the United States.

Türeci, Özlem (born 1967) German researcher and physician, specializing in the development of cancer immunotherapies. Together with her husband, **Ugur Sahin**, she founded Ganymed Pharmaceuticals in 2001, and BioNTech in 2008. She is currently the Medical Director of BioNTech.

Weissman, Drew (born 1959) American scientist, professor of medicine at the University of Pennsylvania (USA), and specialist in messenger RNA therapies.

Printed in the United States
by Baker & Taylor Publisher Services